Edexcel GCSE (9-1)
Biology

Mark Levesley   Sue Kearsey

ALWAYS LEARNING

PEARSON

Contents

Teaching and learning

The **topic reference** tells you which part of the course you are in. 'SB4e' means, 'Separate Science, Biology, unit 4, topic e'.

The **specification reference** allows you to cross reference against the specification criteria so you know which parts you are covering. References that end in B, e.g. B1.14B, are in Biology only, the rest are also in the Combined Science specification criteria.

If you see an **H** icon that means that content will be assessed on the Higher Tier paper only.

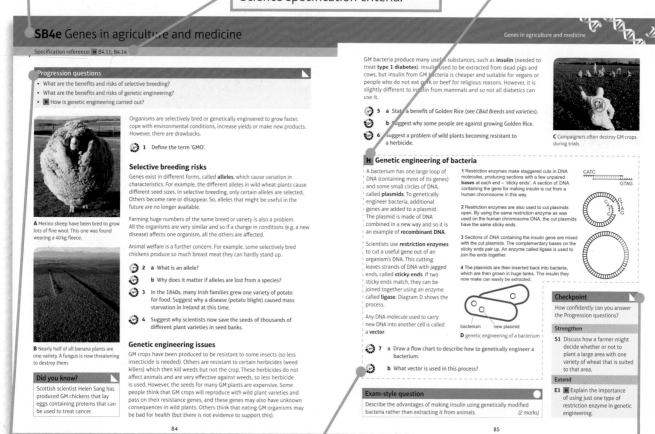

By the end of the topic you should be able to confidently answer the **Progression questions**. Try to answer them before you start and make a note of your answers. Think about what you know already and what more you need to learn.

Each question has been given a **Pearson Step** from 1 to 12. This tells you how difficult the question is. The higher the step the more challenging the question.

When you've worked through the main student book questions, answer the **Progression questions** again and review your own progress. Decide if you need to reinforce your own learning by answering the **Strengthen question**, or apply, analyse and evaluate your learning in new contexts by tackling the **Extend question**.

Paper 1 and Paper 2

SB1 Key Concepts in Biology

The bone-eating snot-flower worm (*Osedax mucofloris*) has no digestive system but still manages to feed on one of the hardest substances produced by vertebrate animals – their bones. These worms are a type of zombie worm, so-called because they have no eyes or mouth, and were discovered in the North Sea in 2005 feeding on a whale skeleton. Enzymes in the 'foot' of the worm cause the production of an acid, which attacks bone and releases lipids and proteins from inside the bone. Enzymes in bacteria on the foot of the worm then digest these large organic molecules into smaller molecules that the worms absorb (using processes such as diffusion).

In this unit you will learn about some of the central ideas in biology, including ideas about cells, microscopy, enzymes, nutrition, diffusion, osmosis and active transport.

The learning journey

Previously you will have learnt at KS3:

- how to use a microscope
- about the differences between cells from different organisms
- how some cells are specialised and adapted to their functions
- how enzymes help to digest food in the digestive system
- how substances can move by diffusion.

In this unit you will learn:

- how developments in microscopy have allowed us to find out more about the sub-cellular structures found in plant, animal and bacterial cells
- about the importance of enzymes in nutrition, growth and development
- how enzymes are affected by pH and temperature and why each enzyme only works on a certain type of molecule
- how to carry out food tests and calorimetry
- how substances are carried by diffusion, osmosis and active transport.

The 'foot' of the worm is buried in the whale bone and contains many bacteria.

SB1a Microscopes

Specification reference: B1.3; B1.4; B1.5

Progression questions

- What determines how good a microscope is at showing small details?
- What has the development of the electron microscope allowed us to do?
- What units are used for very small sizes?

eyepiece lens

focusing wheel – adjusts the focus to make the image clearer

objective lens

specimen holder

A Hooke's microscope

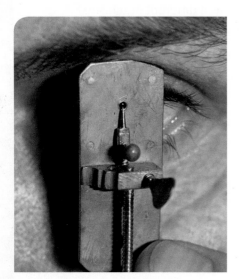

B replica of a van Leeuwenhoek microscope

The most common microscope used today contains two lenses and was invented at the end of the 16th century. Robert Hooke (1635–1703) used a microscope like this to discover cells in 1665.

Hooke's microscope had a **magnification** of about ×30 (it made things appear about 30 times bigger). A person magnified 30 times would be roughly the size of the Statue of Liberty in New York.

1 a A photo of a water flea says it is magnified ×50. What does this mean?

b On the photo, the flea is 5 cm long. Calculate the unmagnified length of the water flea.

To work out a microscope's magnification, you *multiply* the magnifications of its two lenses together. So, the magnification of a microscope with a ×5 **eyepiece lens** and ×10 **objective lens** is:

$$5 \times 10 = \times 50$$

2 A microscope has a ×5 eyepiece lens with ×5, ×15 and ×20 objective lenses. Calculate its three total magnifications.

Hooke's microscope was not very powerful because the glass lenses were of poor quality. Antonie van Leeuwenhoek (1632–1723) found a way of making much better lenses, although they were very small. He used these to construct microscopes with single lenses, which had magnifications of up to ×270. In 1675, he examined a drop of rainwater and was surprised to find tiny organisms, which he called 'animalcules'. Fascinated by his discovery, he searched for 'animalcules' in different places.

3 The top bacterium in photo C is 0.002 mm long in real life. At what magnification is the drawing?

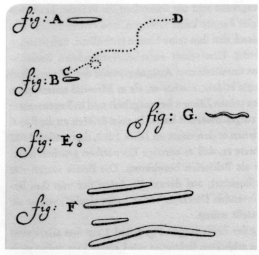

C These are van Leeuwenhoek's drawings of 'animalcules' found in scrapings from his teeth. We call them bacteria.

Did you know?

Van Leeuwenhoek examined his semen and discovered sperm cells.

2

The detail obtained by a microscope also depends on its **resolution**. This is the smallest distance between two points that can still be seen as two points. Van Leeuwenhoek's best microscopes had a resolution of 0.0014 mm. Two points that were 0.0014 mm or further apart could be seen as two points, but two points closer together than this appeared as a single point.

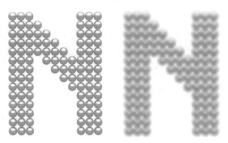

D These images of tiny beads have the same magnification but different resolutions.

 4 Hooke's microscope had a resolution of about 0.002 mm. What does this mean?

With the development of **stains** for specimens, and better lenses and light sources, today's best light microscopes magnify up to ×1500 with resolutions down to about 0.0001 mm.

The electron microscope was invented in the 1930s. Instead of light, beams of electrons pass through a specimen to build up an image. These microscopes can magnify up to ×2 000 000, with resolutions down to 0.0000002 mm. They allow us to see cells with great detail and clarity.

 5 Explain why electron microscope images show more detail than light microscopes.

SI units

The measurements on these pages are in millimetres. Adding the 'milli' prefix to a unit divides it by 1000. One metre (m) contains 1000 millimetres (mm). There are other prefixes that often make numbers easier to understand.

Table E

Prefix	Effect on unit	Example
milli-	÷ 1000	millimetres (mm)
micro-	÷ 1 000 000	micrometres (μm)
nano-	÷ 1 000 000 000	nanometres (nm)
pico-	÷ 1 000 000 000 000	picometres (pm)

× 1000 ↻
× 1000 ↻
× 1000 ↻

↻ ÷ 1000
↻ ÷ 1000
↻ ÷ 1000

 6 Give the highest resolution of electron microscopes in micro-, nano- and picometres.

Exam-style question

State two advantages of using an electron microscope to view cells, rather than a light microscope. *(2 marks)*

SB1b Plant and animal cells

Specification reference: B1.1; B1.4; B1.6

Progression questions

- How are animal cells different to plant cells?
- What do the sub-cellular structures in eukaryotic cells do?
- How can we estimate the sizes of cells and their parts?

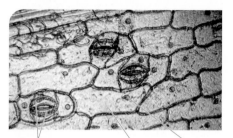

two guard cells (form a stoma in the surface of a leaf)   leaf surface cell   nucleus

A This micrograph ('microscope picture') was taken using Brown's original microscope, of the same cells in which he discovered nuclei (magnification ×67).

As microscopes improved, scientists saw more details inside cells. In 1828, Robert Brown (1773–1858) examined cells from the surface of a leaf and noticed that each cell contained a small, round blob. He called this the **nucleus** (meaning 'inner part' in Latin).

 1 Photo A is at a magnification of 67. State what this means.

Brown wrote a **scientific paper** about his discovery. Matthias Schleiden (1804–1881) read the paper and thought that the nucleus must be the most important part of a plant cell. He mentioned this idea to Theodor Schwann (1810–1882), who then wondered if he could find cells with nuclei in animals. He did. And so the idea of cells being the basic building blocks of all life was born.

A cell with a nucleus is described as **eukaryotic**. We have now discovered many other sub-cellular ('smaller than a cell') structures in eukaryotic cells and worked out what they do.

The **cell membrane** is like a very thin bag. It controls what enters and leaves, and separates one cell from another.

The **cytoplasm** contains a watery jelly and is where most of the cell's activities occur.

One of these blobs is a **mitochondrion** (see photo C). Mitochondria are jelly-bean shaped structures in which **aerobic respiration** occurs. Mitochondria are very difficult to see with a light microscope.

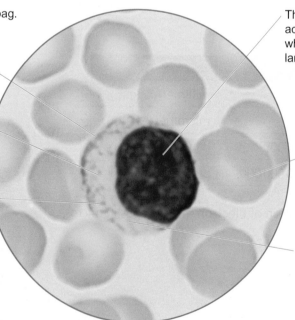

The **nucleus** controls the cell and its activities. Inside it are **chromosomes**, which contain **DNA**. It is especially large in white blood cells.

red blood cell

The cytoplasm also contains tiny round structures called **ribosomes**. These make new proteins for a cell. It is impossible to see them with a light microscope.

B The labelled central cell is a human white blood cell, which has been stained to make its features show up clearly (magnification ×2500).

 2 Draw a table to show the parts of an animal cell and the function of each part.

 3 Estimate the diameter of the labelled red blood cell in photo B. Show your working.

The circular area you see in a light microscope is the **field of view**. If we know its diameter, we can estimate sizes. The diameter of the field of view in photo B is 36 μm. We can imagine that three white blood cells will roughly fit across the field of view. So the cell's diameter is about $\frac{36}{3}$ = 12 μm.

Electron micrographs

Photo C shows many parts inside a white blood cell that you cannot see with a light microscope. However, you still cannot see ribosomes because they are only about 25 nm in diameter.

 4 a Look at photo C. What part has been coloured purple?

 b Use the magnification to estimate the width of the cell.

 5 State the diameter of a ribosome in micrometres.

Scale bars are often shown on micrographs and these are also used to estimate sizes. The scale bar on photo C shows how long 4 µm is at this magnification. About three of these bars could fit across the cell at its widest point; the cell is about 3 × 4 = 12 µm wide.

Plant cells may have some additional structures compared with animal cells, as shown in diagram D.

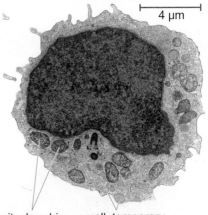

mitochondria   small, temporary vacuoles

C electron micrograph of a white blood cell (magnification ×4200)

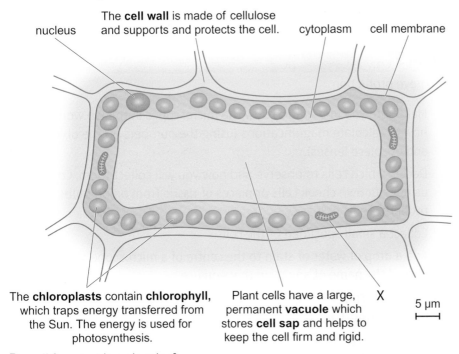

nucleus   The **cell wall** is made of cellulose and supports and protects the cell.   cytoplasm   cell membrane

The **chloroplasts** contain **chlorophyll,** which traps energy transferred from the Sun. The energy is used for photosynthesis.   Plant cells have a large, permanent **vacuole** which stores **cell sap** and helps to keep the cell firm and rigid.   X   5 µm

D a cell from inside a plant leaf

 7 Look at diagram D. What is part X?

 8 Cells on leaf *surfaces* contain vacuoles and carry out aerobic respiration but are not green. Suggest what part they lack. Explain your reasoning.

> ### Did you know?
>
> The pigment in human skin is made in sub-cellular structures called melanosomes.

6 Use the scale bar on photo C to estimate the:

 a width of the nucleus at its widest point

 b length of the longest mitochondrion (coloured red).

> ### Checkpoint
>
> How confidently can you answer the Progression questions?
>
> ### Strengthen
>
> **S1** Draw a plant cell and label its parts, describing what each part does.
>
> ### Extend
>
> **E1** An 'organelle' is a structure inside a cell with a specific function. Compare the organelles found in plant and animal cells.

> ### Exam-style question
>
> Describe the function of chloroplasts in a leaf palisade cell. *(3 marks)*

SB1b Core practical – Using microscopes

Specification reference: B1.6

Aim

Investigate biological specimens using microscopes, including magnification calculations and labelled scientific drawings from observations.

A Hooke's drawing of cork cells, published in 1665 in his book *Micrographia*

One of the first people to examine cells using a microscope was Robert Hooke. He examined bark from a cork oak tree and saw little box shapes. He called them 'cells' because he thought the boxes looked like the small rooms (or cells) found in monasteries at the time. Hooke also realised the importance of making accurate drawings of what he saw to help explain his work to others.

Your task

You are going to make a slide of some plant or animal tissue and examine it using a microscope. You will then make an accurate drawing of one or more of the cells that you see, and add information to help people understand your drawing (e.g. labelling the cell, adding a scale bar).

Method

Wear eye protection.

A Make sure you understand how the microscopes in the lab work, and how to calculate magnifications (using the numbers on the objective and eyepiece lenses).

B Decide which cells to observe and how you will collect them. Consider using your own cheek cells or pieces of tissue from onion bulbs, rhubarb stems or leaves.

C Collect a small specimen of the cells.

D Add a drop of water or stain to the centre of a microscope slide. Record the name of any stain that you use.

E Place your specimen on the drop of water or stain.

F Use a toothpick to slowly lower a coverslip onto the specimen, as shown in diagram B. The coverslip keeps the specimen flat, holds it in place and stops it drying out.

G Examine your specimen under a microscope. Start with the lowest magnification and work up to higher magnifications.

H Draw one or more of the cells that you see and annotate your drawing appropriately.

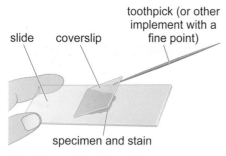

slide   coverslip   toothpick (or other implement with a fine point)

specimen and stain

B Lowering a coverslip slowly and carefully means a slide is less likely to contain air bubbles.

Exam-style questions

1 State the name of the part of a microscope where you would place the slide. *(1 mark)*

2 Photo C shows a light microscope.
   a Give the letter of the part that is an objective lens. *(1 mark)*
   b Give the letter of a part that is used to focus an image. *(1 mark)*

3 State why the lowest power magnification is used when first examining a specimen. *(1 mark)*

4 A microscope is fitted with three objective lenses (of ×2, ×5 and ×10).
   a State what ×2 on a lens means. *(1 mark)*
   b The microscope has a ×7 eyepiece lens. Calculate the possible total magnifications. Show your working. *(3 marks)*

5 Luka has made a slide of some onion tissue. When he examines the specimen with a light microscope, he sees large, thick-walled circles that make it difficult to observe the cells.
   a Give the reason for Luka's observation. *(1 mark)*
   b State how he could prepare a better slide. *(1 mark)*

6 When looking at plant root tissue under a microscope, Jenna notices that about 10 cells fit across the field of view. She calculates the diameter of the field of view as 0.2 mm. Estimate the diameter of one cell. Show your working. *(2 marks)*

7 Photo D shows a certain type of white blood cell called a neutrophil. The image was taken using an electron microscope.
   a State one advantage of using an electron microscope rather than a light microscope. *(1 mark)*
   b Calculate the diameter of the cell to the nearest whole micrometre using the scale bar. *(1 mark)*
   c Give your answer to part b in mm. *(1 mark)*
   d Draw the cell and label the nucleus, cell membrane and cytoplasm. *(2 marks)*

8 Sasha draws a palisade cell from a star anise plant. The cell has a length of 0.45 mm.
   a Sasha's drawing is magnified ×500. Calculate the length of the cell in Sasha's drawing. *(1 mark)*
   b Sasha adds a scale bar to show 0.1 mm. Calculate the length of the scale bar. *(1 mark)*

9 A heart muscle cell is 20 μm wide. It has been drawn 1 cm wide. Calculate the magnification of the drawing. *(2 marks)*

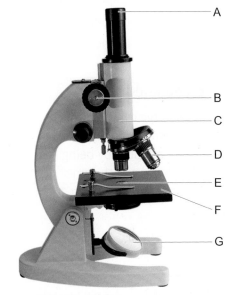

C a light microscope

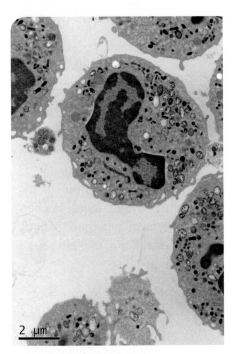

2 μm

D human neutrophils

7

SB1c Specialised cells

Specification reference: B1.2; B1.4; B1.6

Progression questions

- How are some specialised cells adapted to their functions?
- What is the function of a gamete?
- What is the function of cilia?

Did you know?

Human nerve cells (neurones) carry information very quickly. Many are adapted by being extremely long, with some reaching lengths of about 1.4 m.

Specialised cells have a specific function (job). There are about 200 different types of specialised cells in humans. All human cells have the same basic design, but their sizes, shapes and sub-cellular structures can be different so that specialised cells are **adapted** to their functions.

1 List three specialised human cells and state their functions.

Specialised cells for digestion

The cells that line the small intestine absorb small food molecules produced by **digestion**. They are adapted by having membranes with many tiny folds (called **microvilli**). These **adaptations** increase the surface area of the cell. The more area for molecules to be absorbed, the faster absorption happens.

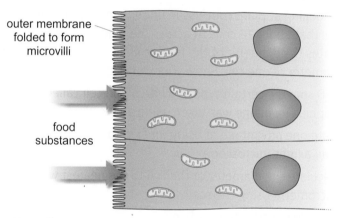

outer membrane folded to form microvilli

food substances

A small intestine cells

2 a Draw a small intestine cell and label its parts.

b These cells are 20 μm long. Add a 10 μm scale bar to your drawing.

c Explain why a cell with microvilli absorbs substances more quickly than one without.

3 Cells called hepatocytes make a lot of a substance called serum albumin. These cells contain many ribosomes. Suggest what type of substance serum albumin is. Explain your reasoning.

4 Nerve cells require a lot of energy. Suggest the adaptation that allows them to get enough energy.

5 a State whether a sperm cell is haploid or diploid.

b Explain why it needs to be like this.

Cells in an organ called the pancreas make **enzymes** needed to digest certain foods in the small intestine. The enzymes are proteins and so these cells are adapted by having a lot of ribosomes.

The wall of the small intestine has muscles to squeeze food along. The muscle cells require a lot of energy and are adapted by having many mitochondria.

Specialised cells for reproduction

During sexual reproduction, two specialised cells (**gametes**) fuse to create a cell that develops into an **embryo**. Human gametes are the **egg cell** and the **sperm cell**.

Most human cell nuclei contain two copies of the 23 different types of chromosome. Gametes contain just *one* copy of each. This means that the cell produced by **fertilisation** has two copies. Cells with two sets of chromosomes are **diploid** and those with one copy of each chromosome are **haploid**.

The cell membrane fuses with the sperm cell membrane. After fertilisation, the cell membrane becomes hard to stop other sperm cells entering.

The jelly coat protects the egg cell. It also hardens after fertilisation, to ensure that only one sperm cell enters the egg cell.

haploid nucleus

The cytoplasm is packed with nutrients, to supply the fertilised egg cell with energy and raw materials for the growth and development of the embryo.

B adaptations of a human female gamete

streamlined shape

The tip of the head contains a small vacuole called the **acrosome.** It contains enzymes that break down the substances in the egg cell's jelly coat. This allows the sperm cell to burrow inside.

nucleus

The tail waves from side to side, allowing the sperm cell to swim.

A large number of mitochondria are arranged in a spiral around the top of the tail, to release lots of energy to power the tail.

cell surface membrane

|— 10 μm —|

C adaptations of a human male gamete

Fertilisation occurs in the **oviducts** of the female reproductive system. Cells in the lining of the oviduct transport egg cells (or the developing embryos after fertilisation) towards the uterus. The oviduct cells are adapted for this function by having hair-like **cilia**. These are like short sperm cell tails and wave from side to side to sweep substances along. Cells that line structures in the body are called **epithelial cells**, and epithelial cells with cilia are called **ciliated epithelial cells**.

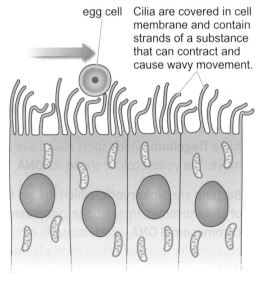

egg cell   Cilia are covered in cell membrane and contain strands of a substance that can contract and cause wavy movement.

D adaptations of oviduct lining cells

 7 Compare and contrast microvilli and cilia.

 8 Explain why an egg cell does not need a tail but a sperm cell does.

 6 a Make a drawing of a human egg cell and label its parts.

b Describe how an egg cell is adapted to prevent more than one sperm cell entering.

 c A human egg cell has a diameter of 0.1 mm. Calculate the magnification of your drawing.

Checkpoint

How confidently can you answer the Progression questions?

Strengthen

S1 List the steps that occur between an egg cell entering an oviduct and it becoming an embryo, and explain how adaptations of specialised cells help each step.

Extend

E1 Explain how both human gametes are adapted to ensure that the cell produced by fertilisation can grow and develop.

Exam-style question

Explain how cells that line the oviduct are adapted to their function of moving the egg cell. *(2 marks)*

SB1e Enzymes and nutrition

Specification reference: B1.12

Progression questions

- What are enzymes made out of?
- What do enzymes do?
- Why are enzymes important for life?

Most animals get substances for energy, growth and development by digesting food inside their bodies. Bacteria, on the other hand, release digestive enzymes into their environments and then absorb digested food into their cells. Starfish use a similar trick for large items of food.

In humans, digestive enzymes turn the large molecules in our food into the smaller subunits they are made of. The digested molecules are then small enough to be absorbed by the small intenstine.

A To eat large items of food, a starfish pushes its stomach out of its mouth and into the food. The stomach surface releases enzymes to break down the food, which can then be absorbed.

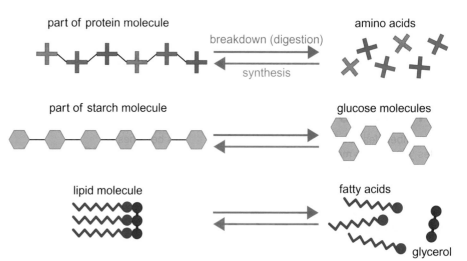

1 Which small molecules make up the following large molecules?

6th   **a** carbohydrates

6th   **b** proteins

6th   **c** lipids

7th   **2** When you chew a piece of starchy bread for a while it starts to taste sweet. Suggest a reason for this.

3 Which monomers make up:

6th   **a** proteins

6th   **b** carbohydrates?

B Large molecules such as complex carbohydrates, proteins and lipids (fats and oils) are built from smaller molecules.

Once the small molecules are absorbed into the body, they can be used to build the larger molecules that are needed in cells and tissues. Building larger molecules from smaller subunits is known as **synthesis**. Complex carbohydrates and proteins are both **polymers** because they are made up of many similar small molecules, or **monomers**, joined in a chain.

The breakdown of large molecules happens incredibly slowly and only if the bonds between the smaller subunits have enough energy to break. Synthesis also happens very slowly, since the subunits rarely collide with enough force or in the right orientation to form a bond. These reactions happen much too slowly to supply all that the body needs to stay alive and be active.

Many reactions can be speeded up using a **catalyst**. In living organisms, the catalysts that speed up breakdown (e.g. digestion) and synthesis reactions are enzymes. So enzymes are **biological catalysts** that increase the rate of reactions. Enzymes are a special group of proteins that are found throughout the body. The substances that enzymes work on are called **substrates**, and the substances that are produced are called **products**.

 4 Define the term 'biological catalyst'.

 5 a Which type of smaller molecule are enzymes built from?

 b Explain your answer.

Enzyme name	Where found	Reaction catalysed
amylase	saliva and small intestine	breaking down starch to small sugars, such as maltose
catalase	most cells, but especially liver cells	breaking down hydrogen peroxide made in many cell reactions to water and oxygen
starch synthase	plant	synthesis of starch from glucose
DNA polymerase	nucleus	synthesis of DNA from its monomers

C examples of enzymes, where they are found and what they do

 6 Name the substrate of amylase, and the products of the reaction it catalyses.

 7 Give two examples of processes that are controlled by enzymes in the human body.

 8 Suggest what will happen in the cells of someone who does not make phenylalanine hydroxylase. Explain your answer.

 9 Sketch a diagram or flowchart to explain how the starfish in photo A absorbs food molecules into its body.

Exam-style question

Explain why the role of enzymes as catalysts in digestion is important for life.

(3 marks)

Did you know?

The heel prick test takes a small amount of blood to test for several factors, including the enzyme phenylalanine hydroxylase. This enzyme catalyses the breakdown of an amino acid called phenylalanine. A few babies are born without the ability to make the enzyme, which can result in nerve and brain damage as they grow older.

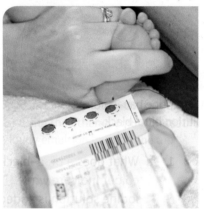

D Babies are given the heel prick test before they are a week old.
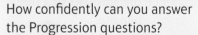

Checkpoint

How confidently can you answer the Progression questions?

Strengthen

S1 Draw a concept map that includes all the important points on these pages. Link words to show how they are related.

Extend

E1 Many bacteria have flexible cell walls made by linking together chains of a polymer. The links are formed in reactions catalysed by an enzyme. Penicillin stops this enzyme from working. Explain how penicillin causes bacteria to be weakened.

Aim

Investigate the use of chemical reagents to identify starch, reducing sugars, proteins and fats.

The lemonade contains sucrose, which is a non-reducing sugar. It does not show a reaction in the Benedict's test.

The energy drink contains glucose and so shows a positive result in the Benedict's test.

A The glucose in the sports 'energy' drink makes it more useful than lemonade as the glucose can be easily absorbed during exercise for use in respiration.

By law, all packaged food and drink must be labelled to show how much fat, sugar, protein and some other substances they contain. This is to help customers make informed choices about what they eat and drink. Every new food or drink that is developed must be tested to produce the information needed for the labelling.

Your task

You will be given a selection of foods. You should plan an investigation to identify whether each food contains starch, reducing sugars, proteins and fats.

Method

c. means 'approximately' or 'about'. Protein concentrations slightly below or above the value will give a similar colour.

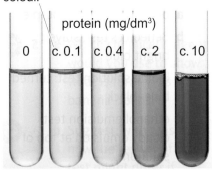

B The results of the biuret test using known concentrations of protein. The tubes can be compared with the results from food samples to give an approximate idea of the amount of protein in the food.

When planning your investigation:

- Decide which test to use for each substance.
- Describe how you will carry out each of the tests.
- Identify how you will know if a test is positive for a particular substance, including the use of control samples.
- Identify any hazards and how you will reduce risk.
- Wear eye protection.

You will need to start by preparing a solution of each food in water.

Some food tests are **qualitative**. This means they show only whether the substance is present or not. This kind of information can answer some questions that we want to ask. For example, some people react badly to some substances and so need to know whether or not they are in a food.

Information may also be **quantitative**, such as the mass of each substance given on a food package label. Some tests, such as the biuret test, can be **semi-quantitative**, meaning that they give some information about value, often in terms of 'little', 'some' or 'lots', or approximate values.

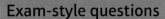

Exam-style questions

1 State which substance is identified in each of the following tests.

   a iodine solution test   **c** ethanol emulsion test

   b biuret test   **d** Benedict's test   *(4 marks)*

2 **a** Name all the apparatus that you would need to carry out each test in question 1.   *(4 marks)*

   b Describe how to carry out each test in question 1.   *(4 marks)*

3 Describe how you would identify any hazards for each of the tests.

   (1 mark)

4 Describe a positive result and a negative result for each of the tests.   *(4 marks)*

5 Predict the result of each of the following tests. Give a reason for each answer.

   a iodine test on cooked rice   *(2 marks)*

   b Benedict's test on egg white   *(2 marks)*

   c biuret test on egg white   *(2 marks)*

   d ethanol emulsion test on cheese   *(2 marks)*

6 Table C shows the results of some tests on several foods. Use the results to identify which substances each food contains.   *(4 marks)*

Food	Iodine test	Biuret test	Ethanol test	Benedict's test
biscuit	blue–black	blue–purple	white emulsion	red
cauliflower	yellow–orange	light blue	no layer	orange
egg yolk	yellow–orange	purple	white emulsion	blue
fat-free milk	yellow–orange	purple	no layer	green

C

7 **a** Put the foods in the table above in order, from lowest quantities of reducing sugars to highest quantities of reducing sugars.   *(1 mark)*

   b Suggest an explanation for why it is possible to do this for foods containing reducing sugars but not for foods containing starch.   *(2 marks)*

8 Fats or oils can also be tested for by mixing a solution of the food with Sudan III stain. Any fat or oil dissolves in the bright red stain, and floats to the surface of the solution. This gives a bright red layer at the top, rather than redness generally dispersed through the solution.

   a Describe a control that could be used with the Sudan III test. *(1 mark)*

   b Explain why using a control would improve any conclusion drawn from the results.   *(2 marks)*

Bacteria living on your body cause body odour. The smelly substances they produce are released into the air and reach our noses.

Smells spread by **diffusion**. Particles in gases and liquids are constantly moving past each other in random directions. This causes an overall movement of particles from where there are more of them (a higher **concentration**) to where there are fewer (a lower concentration).

A difference between two concentrations forms a **concentration gradient**. Particles diffuse *down* a concentration gradient. The bigger the difference between concentrations, the steeper the concentration gradient and the faster diffusion occurs.

A an experiment to assess body odour

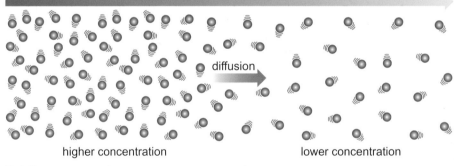

The number of particles *decreases* as you go *down* a concentration gradient.

higher concentration   lower concentration

B diffusion occurs down a concentration gradient

Diffusion allows small molecules (such as oxygen and carbon dioxide) to move into and out of cells.

Osmosis

A membrane that allows some molecules through it and not others is **partially permeable** (or **semi-permeable**).

Cell membranes are semi-permeable and trap large soluble molecules inside cells, but water molecules can diffuse through the membrane. If there are more water molecules in a certain volume on one side of the membrane than the other, there will be an overall movement of water molecules from the side where there are more water molecules (a more dilute **solute** concentration) to the side where there are fewer water molecules (a more concentrated solution of solute). This diffusion of small molecules of a **solvent**, such as water, through a semi-permeable membrane is called **osmosis**. The overall movement of solvent molecules will stop when the concentration of solutes is the same on both sides of a membrane.

 1 Explain why smells spread.

 2 a A dish of perfume is put at the front of a lab. Describe the perfume's concentration gradient after 5 minutes.

b Describe the overall movement of the perfume molecules.

 3 Muscle cells in the leg use up oxygen but are surrounded by a fluid containing a lot of oxygen. Explain why oxygen moves into the cells.

4 **a** In diagram C, in which direction will water flow, X to Y or Y to X?

 b Explain why this flow occurs.

 5 Red blood cells contain many solute molecules. Explain why red blood cells burst if put in pure water.

Osmosis can cause tissues to gain or lose mass. To calculate the mass change:

- work out the difference between the mass of tissue at the start and at the end (final mass – initial mass)
- divide this difference by the initial mass
- multiply by 100.

So, percentage change in mass = $\dfrac{\text{(final mass – initial mass)}}{\text{initial mass}} \times 100$

A negative answer is a percentage *loss* in mass.

 6 An 8 g piece of potato is left in water for an hour. Its mass becomes 8.5 g. Calculate the percentage change in mass.

Active transport

Cells may need to transport molecules *against* a concentration gradient or transport molecules that are too big to diffuse through the cell membrane. They can do this using **active transport**.

This process is carried out by transport proteins in cell membranes. The transport proteins capture certain molecules and carry them across the cell membrane. This is an active process and so requires energy. Osmosis and diffusion are **passive** processes, so do not require an input of energy.

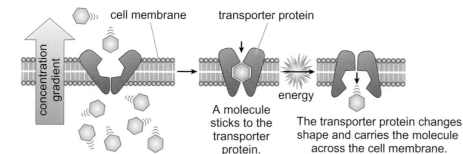

D active transport

 8 Explain how cells that carry out a lot of active transport would be adapted to their function.

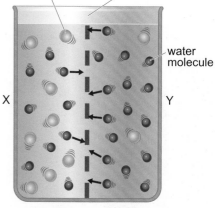

soluble molecule that is too large to pass through the membrane (e.g. sucrose)

partially permeable membrane allows molecules to pass through if they are small enough

water molecule

X Y

more concentrated solution

more dilute solution

C In osmosis, a solvent flows from a dilute solution of a solute to a more concentrated one.

 7 Look at diagram D. Explain why active transport is needed to move the molecules.

Checkpoint

How confidently can you answer the Progression questions?

Strengthen

S1 A small number of sugar molecules are in your small intestine. Describe how they will be absorbed into cells in the small intestine and why they need to be absorbed in this way.

Extend

E1 Sorbitol is a sweet-tasting substance that is not broken down or absorbed by the body. It is used in some sugar-free sweets. Explain why eating too many of these sweets can cause diarrhoea.

Exam-style question

Explain why a slice of potato will decrease in mass if it is placed in a concentrated sugar solution. *(2 marks)*

SB1i Core practical – Osmosis in potato slices

Specification reference: B1.16

Aim

Investigate osmosis in potatoes.

A The tiny white crystals on this cord grass are salt released from glands on the surface of the leaves.

Most land plants die if flooded with sea water. Usually the concentration of mineral salts in the soil is less than inside the roots, so water moves into the root from the soil by osmosis. If the soil contains a high concentration of salts then osmosis occurs out of the root, into the soil. Without sufficient water, plants die. Some plants, called halophytes, are adapted to live in salty areas. Their roots take in large amounts of salts from the soil, which helps osmosis to continue from the soil into the plant. Halophytes get rid of the extra salt they absorb in various ways.

Your task

You are going to measure osmosis in plant tissue, by comparing the mass of the tissue before and after soaking in sucrose solution. Sucrose is used because the molecules are too large to diffuse through cell membranes. The change in the mass of the tissue shows how much water is absorbed or lost. You can work out the solute concentration of the plant tissue from repeating the experiment with solutions of different concentrations.

When comparing different pieces of tissue, remember to calculate the percentage change in mass of each piece.

Method

A Label a separate boiling tube with the sucrose concentration of each solution you will test. Place all the tubes in a rack.

B Cut similar-sized pieces of potato, enough for one per tube. (Make sure they fit in a tube.)

C Blot each potato piece dry, measure and record its mass, and put it in an empty tube.

D Fill each tube with the solution of the appropriate concentration. Ensure you cover the potato with the solution.

E After at least 15 minutes, remove each potato piece and blot it dry. Measure and record its mass again.

Exam-style questions

1   Define the term 'osmosis'.   *(1 mark)*

2   Use the idea of osmosis to explain why most plants in salty soil would have problems absorbing water through their roots.   *(2 marks)*

3   Explain how halophyte roots are adapted to help them absorb water from salty soil.   *(2 marks)*

4   **a**   State which variables have been kept the same in the method on the previous page.   *(1 mark)*

   b   Explain the importance of controlling these variables.   *(2 marks)*

5   Write an equipment list for the method shown on the previous page.   *(2 marks)*

6   Explain how you would work out a suitable range of sucrose concentrations for the solutions in the potato experiment.   *(2 marks)*

7   Give a reason why you should calculate percentage change in mass when comparing results.   *(1 mark)*

8   State one way to improve the method for the potato experiment so that you could be more certain of the results.   *(1 mark)*

9   Table B shows the results from an experiment similar to the one described in the method.

   a   For each solution, calculate the gain or loss in mass of the potato piece.   *(2 marks)*

   b   For each solution, calculate the percentage change in mass of the potato.   *(2 marks)*

   c   Give a reason for the result from tube A.   *(1 mark)*

   d   Explain the results from tubes B–D.   *(2 marks)*

   e   Use the results to give the possible solute concentration of potato tissue, giving a reason for your answer.   *(2 marks)*

   f   Describe how the method could be adapted to give a more accurate answer to part **e**.   *(1 mark)*

10 Graph C shows the results of an experiment comparing osmosis in tissue from a halophyte plant and a potato in the same solution.

   a   Identify, with a reason, which tissue lost water fastest over the first five minutes.   *(2 marks)*

   b   Explain why it lost water faster than the other tissue.   *(2 marks)*

   c   Calculate the average rate of change in mass over the first four minutes for the potato.   *(1 mark)*

Tube	A	B	C	D
Sucrose concentration (%)	0	10	30	50
Mass of potato at start (g)	4.81	5.22	4.94	4.86
Mass of potato at end (g)	4.90	4.96	4.39	3.69

B the effect of sucrose concentration on the change in mass of potato (concentration 0 is distilled water)

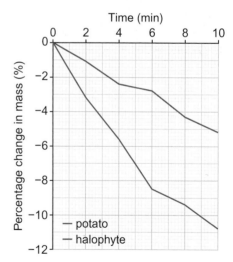

C change in mass of a halophyte plant and potato in the same solution

Arginase

Unused amino acids are broken down into a waste substance called urea. One enzyme involved in the process is called arginase, which catalyses this reaction:

arginine + water → ornithine + urea

In an experiment, the activity of arginase was tested at different pHs. The graph below shows the results. Explain the effect of pH on arginase.

(6 marks)

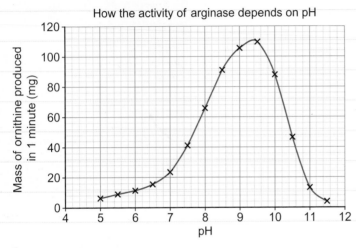

How the activity of arginase depends on pH

A

. . . .

Student answer

The pH affects arginine [1] and makes it make the reaction faster or slower. It shows that the enzyme is best in high pH [2].

[1] The name of the enzyme is arginase (the substrate is arginine). It is very important to use the correct scientific names.

[2] The answer correctly says that the enzyme is more active at higher pHs.

. . . .

Verdict

This is a weak answer. It only contains one correct fact and there is no explanation about why pH has an effect on the enzyme.

The key word in the question is 'explain'. The answer could be improved by explaining the link between the shape of the graph and how the enzyme works (i.e. the enzyme works best at the optimum pH because this is when the active site has the best shape for the substrate to fit into it). A really good answer would also have used data from the graph (e.g. pointing out that the optimum pH is 9.4).

Exam tip

It may help to note down some key facts that you know about the topic before you write your answer. Cross out your notes when you have finished writing.

Paper 1

SB2 Cells and Control

The blue whale (*Balaenoptera musculus*) is the largest animal ever to have lived on Earth – larger than the biggest dinosaurs. Blue whales grow to over 30 m in length and have masses of over 150 tonnes. The mass of a whale is, however, not just made up of trillions of different whale cells but also thousands of other organisms (such as 'whale lice' that live on their bodies, and tonnes of bacteria in their digestive systems).

In this unit you will discover how plants and animals develop from single cells the size of full stops to become complex organisms made of many different types of cells, which all need to be controlled and coordinated.

The learning journey

Previously you will have learnt at KS3:

* that cells divide
* about the structure of plant and animal cells (including the chromosomes in their nuclei)
* that your nervous system helps to coordinate your actions.

In this unit you will learn:

* about mitosis and its importance in growth, repair and asexual reproduction
* how cells become specialised, and the importance of stem cells
* to identify different specialised cells in the nervous system and explain how the system works
* how the eye works, and how some eye problems are corrected.

SB2a Mitosis

Specification reference: B2.1; B2.2; B2.3; B2.4

Progression questions

- Why is mitosis important?
- What happens in the different stages of mitosis?
- How do cancer tumours occur?

Every living thing needs to be able to grow and to repair itself in order to stay alive. In organisms that are made of many cells (**multicellular organisms**) the processes of growth and repair require new cells. These are produced in a process called the **cell cycle**.

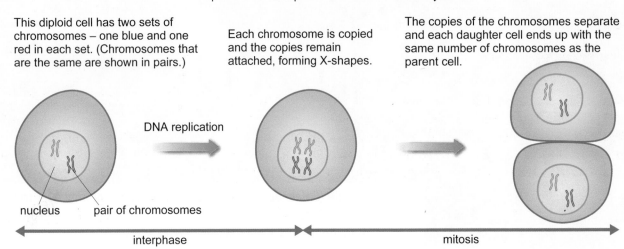

This diploid cell has two sets of chromosomes – one blue and one red in each set. (Chromosomes that are the same are shown in pairs.)

Each chromosome is copied and the copies remain attached, forming X-shapes.

The copies of the chromosomes separate and each daughter cell ends up with the same number of chromosomes as the parent cell.

DNA replication

nucleus   pair of chromosomes

interphase   mitosis

A During the cell cycle two identical daughter cells are formed from a parent cell.

The nuclei of human body cells contain two copies of each of 23 types of chromosome, making 46 in all. Cells with two copies of each chromosome (two sets of chromosomes) are **diploid**. Gametes (sex cells) contain one copy of each type of chromosome and are **haploid**.

There are two phases in the cell cycle, the first of which is **interphase**. In this phase the cell makes extra sub-cellular cell parts (e.g. mitochondria). **DNA replication** (copying) also occurs, to make copies of all the chromosomes. The copies of the chromosomes stay attached to each another, making the chromosomes look like Xs.

The next phase of the cell cycle is cell division or **mitosis**. The cell splits to form two **daughter cells**, which are both identical to the parent cell. Mitosis occurs in a series of continuous stages, shown in diagram B.

6th **1** Give the name of one diploid and one haploid cell in the body of a mammal.

7th **2** Alligators have eight types of chromosome. How many chromosomes are in a diploid alligator cell?

7th **3** Why is DNA replication important in interphase?

8th **4** Why must the number of mitochondria double in a cell during interphase?

7th **5** Draw a table to show what happens in each stage of mitosis.

Did you know?

Damaged human organs cannot regrow … apart from the liver. Liver transplants are often done using part of a liver because the transplanted piece of liver grows by mitosis to form a full-sized liver. The liver pieces for transplants can be taken from living donors because livers grow back.

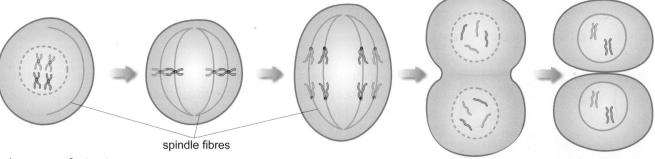

In **prophase** the nucleus starts to break down and **spindle fibres** appear.

By the end of **metaphase**, the chromosomes are lined up on the spindle fibres across the middle of the cell.

The chromosome copies are separated and moved to either end of the cell on the spindle fibres. This is **anaphase**.

In **telophase** a membrane forms around each set of chromosomes to form nuclei.

A cell surface membrane forms to separate the two cells during **cytokinesis.** Cell walls form in plant cells.

spindle fibres

B the stages of mitosis

Asexual reproduction

Some organisms can reproduce using just one parent. This **asexual reproduction** produces offspring that are **clones**, which means that their cells have the same chromosomes as the parent (they are genetically identical). So, asexual reproduction relies on mitosis. Strawberry plants, for example, reproduce asexually using stems that grow along the ground, called runners, and potatoes use tubers. Some animals, such as aphids, can also reproduce asexually. Asexual reproduction is much faster than sexual reproduction because organisms do not need others for reproduction. However, sexual reproduction produces variation and asexual reproduction does not.

C Asexual reproduction is rare in larger animals, but female Komodo dragons can reproduce asexually.

D plantlets produced by asexual reproduction growing on the leaf margins of *Kalanchöe* plant

Growth of cancer tumours

Normal cells only divide when they need to. Changes in cells can, however, sometimes turn them into **cancer cells**, which means that they undergo uncontrollable cell division. This rapid cell division produces growing lumps of cells called **tumours** that can damage the body and can result in death.

6 Why is each plantlet in photo D a clone?

7 Daughter cells produced by mitosis are said to be 'genetically identical' to the parent cell. Explain what this means.

8 Why does asexual reproduction rely on mitosis?

9 A rose plant has a 'crown gall tumour' on its stem.

   a What would you expect this to look like?

   b Explain how this occurs.

Checkpoint

How confidently can you answer the Progression questions?

Strengthen

S1 Draw a flow chart to show mitosis.

Extend

E1 Explain why mitosis is important for the reproduction of organisms if there are very few members of the opposite sex in an area.

Exam-style question

Explain why sperm cells cannot be produced using the cell cycle. *(2 marks)*

SB2b Growth in animals

Specification reference: B2.5; B2.6; B2.7

Progression questions

- Which processes in animals result in growth and development?
- How are percentile charts used to monitor growth in children?
- Why is cell differentiation important in animals?

caterpillar
preparing to pupate
pupa
adult

A During the pupal stage, a caterpillar digests itself! Only some cells remain, but using cell division and differentiation these cells produce all the specialised cells in the adult butterfly.

Growth is an increase in size as a result of an increase in number or size of cells. The number of cells increases due to cell division by mitosis. Growth can be recorded by taking measurements over time, such as of length or mass.

1 a Suggest how you could measure the growth of a kitten.

b Explain your answer.

2 a Your mass increases when you take in food and drink. Is this an example of growth?

b Explain your answer.

The growth of human babies is regularly checked by measuring them, including mass and length. The measurements are checked on charts to show how well a baby is growing compared to others at the same age.

These charts were created by measuring a very large number of babies. The measurements were divided into 100 groups. When divided like this we can find out what percentage of readings are below a certain value, or **percentile**. For example, 25 per cent of babies will have masses below the 25th percentile line, whereas 75 per cent of babies will be below the 75th percentile line. So, if the 25th percentile for an 8-month-old baby's mass is 8 kg then 25 per cent of 8-month-old babies have a mass below this value.

The curved lines (see graph B) show the rate of growth of a baby who stays at exactly the same percentile within the population. Most babies don't grow at the same rate all the time, so plotting their mass helps to identify whether they are growing normally. Although rate of growth may vary from week to week, a baby should remain near the same percentile curve as it gets older.

The 50th percentile curve shows the growth of a baby of the median (average) size of the population. Half (50%) of all babies will have a mass above this curve and half equal to or below the curve.

B Percentile growth curves for UK baby boys from 2 weeks to 1 year, for mass. The red line that has been plotted on the curves shows the growth of one baby.

Did you know?

Growth is not charted for the first two weeks of life because babies often lose weight as they adjust to feeding from the breast or bottle rather than getting their nutrients from the placenta.

Cell differentiation

Although all animals develop from a single cell, not all the cells in their bodies are the same. Cells produced by mitosis are the same as the cell from which they were formed. However, the new cells may then change in different ways, so they become specialised for different functions. The process that changes less specialised cells into more specialised ones is called **differentiation**.

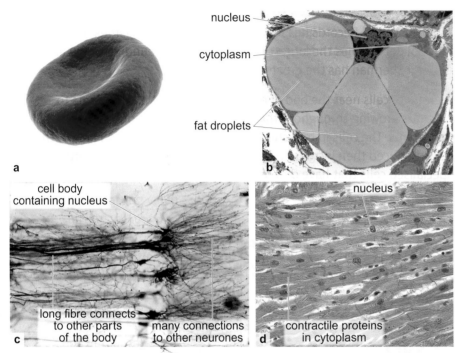

a

b

nucleus

cytoplasm

fat droplets

cell body
containing nucleus

nucleus

long fibre connects
to other parts
of the body

many connections
to other neurones

contractile proteins
in cytoplasm

c

d

C Here are some specialised human cells. (a, top left) A red blood cell has no nucleus, allowing more space for red haemoglobin molecules (which carry oxygen). It also has a large surface area (allowing oxygen to diffuse in and out more quickly). (b, top right) The cytoplasm of fat cells is filled with large fat droplets. The fat is stored until the body needs energy. (c, bottom left) Nerve cells (neurones) have a long fibre that carries electrical impulses around the body and many connections to other neurones. (d, bottom right) Muscle cells contain special contractile proteins that can shorten the cell.

5 a Describe a special feature of a fat cell.

    b Explain how a red blood cell is specialised for its function of carrying oxygen around the body.

6 a Describe two kinds of specialised cells you would expect to find in a butterfly.

    b Explain your choices and predict the adaptations that the cells have.

Exam-style question

Explain why percentile curves are used to measure the growth of babies.

(2 marks)

3 Look at graph B.

    a What is the value of the 50th percentile for mass in a 6-month-old baby boy?

    b How much should the mass of a baby boy in the 50th percentile increase between 3 and 9 months of age?

4 Does the baby plotted on graph B (the red line) show healthy growth? Explain your answer.

Checkpoint

How confidently can you answer the Progression questions?

Strengthen

S1 Describe how a single fertilised human egg cell develops into the billions of different cells in a human adult.

Extend

E1 What are the advantages and disadvantages of using percentile curves to assess the growth and development of a young baby?

SB2d Stem cells

Specification reference: B2.8; B2.9

Progression questions

- Where are stem cells found?
- What is the function of stem cells?
- What are the advantages and risks of using stem cells in medicine?

A Geckos are reptiles that are able to grow a whole new tail if it is cut off. Regrowth of many different tissues to make a new organ is rare in adult animals.

Cells that can divide repeatedly over a long period of time to produce cells that then differentiate are called **stem cells**. In plants, these cells are found in meristems (and are sometimes called **meristem cells**).

Plant stem cells are usually able to produce any kind of specialised cell throughout the life of the plant. This is not true for most animals, especially vertebrates.

1 a State where you can find stem cells in two different plant organs.

b Describe the function of the stem cells in these two plant organs.

2 Stem cells are unspecialised cells. Define the term 'unspecialised'.

Animals start life as a fertilised egg cell, which then divides to form an embryo. The cells of an early-stage embryo are **embryonic stem cells** that can produce any type of specialised cell. As the cells continue to divide, the embryo starts to develop different areas that will become the different organs. The stem cells in these areas become more limited in the types of specialised cell they can produce.

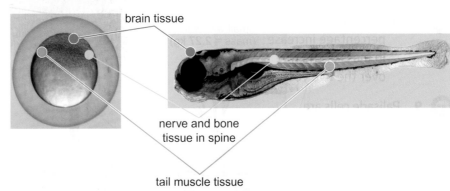

B Stem cells in different parts of a zebrafish embryo (on the left) form different tissues as it develops into a young fish.

By the time the young animal is fully developed, the stem cells can usually only produce the type of specialised cell that is in the tissue around them. These are called **adult stem cells** (even if they are in a young animal). The adult stem cells in human tissues allow the tissues to grow and to replace old or damaged cells.

Did you know?

Zebrafish are able to regenerate many parts of their body using stem cells. They can replace their fins, skin, heart tissue, and even brain cells when they are young. This has made them the most important organism in stem cell research.

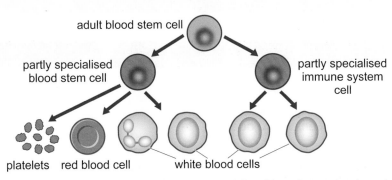

platelets   red blood cell   white blood cells

C Blood stem cells are found in marrow in the middle of long bones (such as the femur). They continue to divide throughout life to produce new blood cells.

Stem cells offer a way of treating many different diseases caused by damaged cells. The first successful human bone marrow transplant was carried out in the late 1950s. Healthy bone marrow from one identical twin was given to the other twin who had a blood disease. Since then, scientists have studied other ways to use adult and embryonic stem cells to treat diseases such as type I diabetes, or to replace damaged cells. This is done by stimulating stem cells to make them produce the specialised cells that are needed and then injecting them into the places they are needed.

D Young zebrafish are transparent. This makes them useful for studying how vertebrate stem cells work and finding out which drugs or treatments might affect stem cells in the body. Successful treatments might then be tried on humans.

There are problems with using stem cells, which scientists are still trying to solve. For example, if stem cells continue to divide inside the body after they have replaced damaged cells, they can cause **cancer**. Another problem is that stem cells from one person are often killed by the immune system of other people that they are put into. This is called **rejection**.

 7 Describe two risks of using stem cells to treat disease.

 3 What is the function of stem cells in bone marrow?

4 Explain why blood stem cells only produce blood cells.

5 Compare adult and embryonic stem cells in terms of what they can do, and their functions.

6 Explain how stem cells could be used to help repair damaged heart muscle cells in someone who has had a heart attack.

Checkpoint

How confidently can you answer the Progression questions?

Strengthen

S1 a Describe the functions of the different kinds of stem cell in animals and plants.

   b Describe one benefit and one risk of using stem cells in medicine.

Extend

E1 In 2014, scientists studying zebrafish discovered that 'buddy' cells are needed to help one type of stem cell become blood stem cells.

   a Suggest how this research could lead to new treatments for people with diseases.

   b Suggest what risks must be overcome before these treatments can be given to patients.

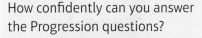

Exam-style question

Describe the role of meristems in plant growth.   *(2 marks)*

SB2e The brain

Specification reference: B2.10B

Progression questions

- What are the brain and spinal cord made of?
- What are the main parts of the brain?
- What do some of the different parts of the brain and spinal cord do?

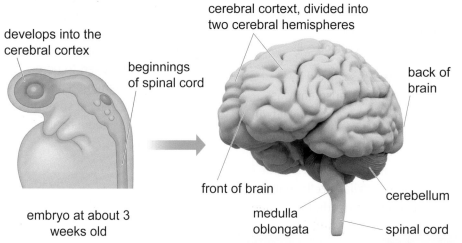

A brain development and some of the major structures of the brain

Labels: develops into the cerebral cortex; beginnings of spinal cord; cerebral cortext, divided into two cerebral hemispheres; back of brain; front of brain; medulla oblongata; cerebellum; spinal cord; embryo at about 3 weeks old

Embryonic stem cells in a human embryo divide to produce more and more stem cells. Once an embryo is three weeks old the stem cells in the brain area start to differentiate to produce **neurones** (nerve cells), which make up most of the brain. An adult brain has about 86 billion neurones, which interconnect with one another and other parts of the body to process information and control the body.

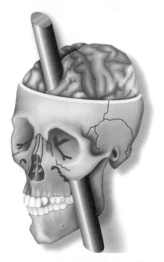

B the position of the metal rod in Gage's brain

 1 State the positions of the cerebral cortex, cerebellum and medulla oblongata in the brain.

 2 Describe how most of the cells in the brain develop.

Cerebral cortex

The **cerebral cortex** makes up 80 per cent of the brain. Its wrinkled surface led the Ancient Greek thinker Aristotle (384–322 BCE) to believe that it acted like a radiator, cooling the body. Modern ideas about its function started to develop thanks to an accident on an American railway line in 1848.

A railway worker called Phineas Gage was checking that a stick of dynamite was properly positioned in a hole, by prodding it with a long metal rod. The dynamite exploded, firing the rod upwards through his head. Remarkably Gage was unconscious for only a few moments before being able to walk and talk normally.

Before the accident, Gage had been hard-working and friendly. Afterwards he became lazy and bad-tempered. Gage's doctor concluded that the front of the cerebral cortex is involved in controlling personality.

Our understanding of the cerebral cortex has developed greatly since this time. We now know that it is used for most of our senses, language, memory, behaviour and consciousness (our inner thoughts and feelings). It is divided into two **cerebral hemispheres**, each with slightly different functions. The right hemisphere communicates with the left side of the body and vice versa.

Did you know?

Underneath each cerebral hemisphere is a structure called the hippocampus. This part of the brain has been found to be bigger in taxi drivers, suggesting that this structure is involved in memory and navigation.

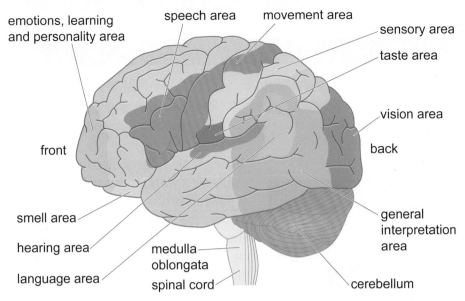

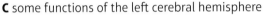

C some functions of the left cerebral hemisphere

Cerebellum

At the base of the brain is the **cerebellum**. It is divided into two halves and controls balance and posture. It also coordinates the timing and fine control of muscle activity, making sure that movements are smooth. Many musicians have developed changes in the cerebellum, including an increase in its size.

5 Suggest why a professional pianist might have a cerebellum that is larger than usual.

6 Use the information in the text to suggest why drinking alcohol might affect a person's balance.

Medulla oblongata

The **medulla oblongata** controls your heart rate and your breathing rate. It is also responsible for **reflexes** such as vomiting, sneezing and swallowing.

The mass of neurones that make up the medulla oblongata connect the brain to the **spinal cord**. The spinal cord is about the width of a finger and consists of many **nerves** (bundles of neurones). These carry information between the brain and the rest of the body.

7 Information is carried from the right cerebral hemisphere to the left hand. Describe the route that the information takes.

8 Suggest why the medulla oblongata is sometimes referred to as the 'automatic pilot' of the brain.

Exam-style question

Some people with ataxia tend to walk with a jerky movement. Explain what part of the brain is likely to be affected by ataxia. *(2 marks)*

3 Explain why Gage's doctor concluded that the front of the cerebral cortex controls personality.

4 Selma shuts her left eye to use a telescope. Explain which area of her brain is most active.

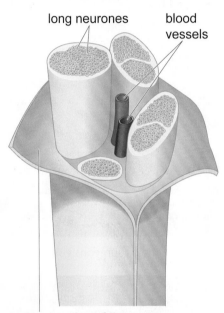

D structure of a nerve

Checkpoint

How confidently can you answer the Progression questions?

Strengthen

S1 Draw a table to summarise the structures and function(s) of the major parts of the brain.

Extend

E1 Jordan sets out on a run through the streets near where he lives. Explain how his brain helps him to complete his run.

SB2f Brain and spinal cord problems

Specification reference: **H** B2.11B; **H** B2.12B

Progression questions

- **H** How is brain function investigated?
- **H** How are spinal injuries treated?
- **H** How are brain tumours treated?

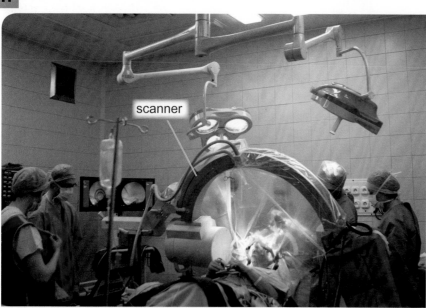

scanner

A A scanner helps surgeons guide instruments during brain surgery.

During brain surgery, electrodes can apply electrical currents to the brain. If patients are awake, they can be asked to describe what they feel. A current may also make a patient do something or stop an action occurring. This allows the functions of brain parts to be investigated.

8th **1** During brain surgery, an electrode causes a patient to experience an acid taste. Where was the electrode placed?

Did you know?

Patients cannot feel brain surgery because the brain has no 'pain receptors'.

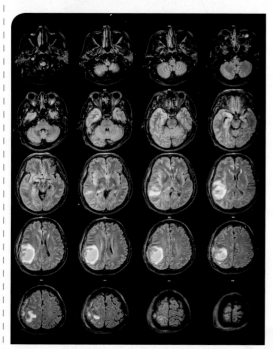

B CT 'slices' through the brain. Denser materials absorb more X-rays, causing whiter areas. The very white area is a brain tumour.

Scanning

Scanning allows scientists to look deeper into the brain than surgery does. It also allows the study of a healthy individual without the risk of damaging the brain.

A **CT scan** shows the shapes of structures in the brain. An X-ray beam moves in a circle around the head, and detectors measure the absorption of the X-rays. A computer uses this information to build up a view of the inside of the body as a series of 'slices'. Differences in the shapes in the brain can be linked to differences in the way people think and act, suggesting the functions of those parts.

7th **2** State two advantages of CT scanning to investigate brain functions compared with using electrodes during brain surgery.

3 The CT scan in photo B is from a patient who had blurred vision in one eye.

9th **a** Describe how the CT scan allowed the cause to be found.

8th **b** Explain which eye was affected.

H

A **PET scan** shows brain activity. The patient is injected with **radioactive** glucose. More active cells take in more glucose than less active ones (for respiration). The radioactive atoms cause **gamma rays**, which the scanner detects. More gamma rays come from parts containing more active cells.

Carrying out activities during a PET scan causes specific areas of the brain to become more active (as shown in photo C).

4 State and explain which type of scanner is being used in photo A.

5 Explain the pattern for 'looking' on the PET scan shown in photo C.

Spinal cord damage

Damage to the spinal cord reduces the flow of information between the brain and parts of the body. Nerve damage in the lower spinal cord can cause loss of feeling in, and use of, the legs. Damage in the neck can cause **quadriplegia** (loss of use of both arms and legs).

There are no adult stem cells that can differentiate into neurones in the spinal cord, and so new neurones cannot be made to repair damage. Wires can be used to electrically stimulate nerves and muscles below the damage, but patients do not regain full movement or feeling. However, treatments using stem cell injections are being developed.

6 Explain why the man in photo D was using a machine to operate his lungs.

7 Suggest how stem cells could be used to treat spinal cord injuries.

Brain tumours

Cancer cells often divide rapidly to form a **tumour** (a lump). A brain tumour may squash parts of the brain and stop them working. Tumours can be cut out or the cells can be killed using **radiotherapy** (high-energy X-ray beams) and **chemotherapy** (injecting drugs that kill actively dividing cells). All these methods can damage the body and brain, and chemotherapy may not work due to the **blood–brain barrier** – a natural filter that only allows certain substances to get from the blood into the brain (mainly due to cells in the capillary walls in the brain fitting together very closely).

8 a Explain why chemotherapy can kill brain tumour cells.

   b Describe two problems of using chemotherapy to kill brain tumour cells.

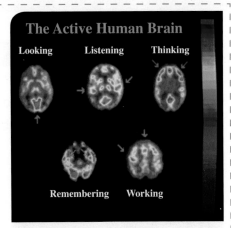

C PET scans allow scientists to match activities with certain areas of the brain.

D This man broke his neck falling off a horse. The machinery on this wheelchair operates the lungs.

Checkpoint

How confidently can you answer the Progression questions?

Strengthen

S1 Karen's spinal cord was cut in an accident. Explain why the damage means that she will never regain full movement or feeling.

Extend

E1 Explain the advantage of combining CT and PET scans to investigate the brain.

SB2g The nervous system

Specification reference: B2.13

Progression questions

- What is the nervous system?
- How does the nervous system allow the body to respond to stimuli?
- How is a sensory neurone adapted to its function?

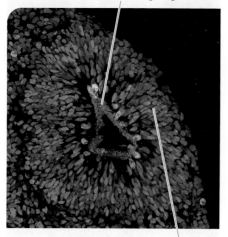

stem cells undergoing mitosis

cells differentiating into nerve cells (neurones)

A This clump of living brain cells was produced from stem cells and grown in the lab. These clumps can be up to 5 mm in diameter.

To study how a human brain works, it is useful for scientists to experiment on living brain tissue. This cannot be done with living people, but in 2015 scientists from Stanford University in the USA managed to grow brain tissue in their lab by using stem cell techniques.

The brain and **spinal cord** form the **central nervous system** (**CNS**), which controls your body. Nerves make up the rest of the **nervous system**. This organ system allows all the parts of your body to communicate, using electrical signals called **impulses**.

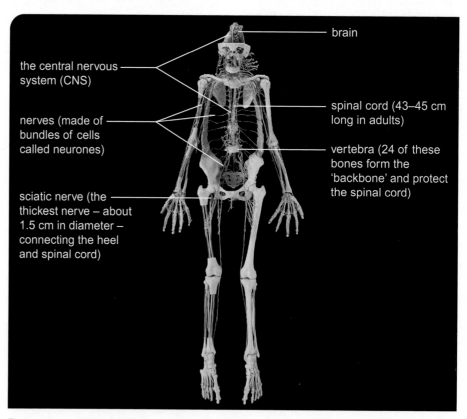

brain

the central nervous system (CNS)

spinal cord (43–45 cm long in adults)

nerves (made of bundles of cells called neurones)

vertebra (24 of these bones form the 'backbone' and protect the spinal cord)

sciatic nerve (the thickest nerve – about 1.5 cm in diameter – connecting the heel and spinal cord)

B dissection of the human nervous system, including some of the thicker nerves

 1 Name the organs in the central nervous system.

Anything your body is sensitive to, including changes inside your body and in your surroundings, is called a **stimulus**. **Sense organs** (such as eyes, ears and skin) contain **receptor cells** that detect stimuli. For example, skin contains receptor cells that detect the stimulus of temperature change.

Receptor cells create impulses, which usually travel to the brain. The brain then processes this information and can send impulses to other parts of the body to cause something to happen (a **response**).

Did you know?

Lined up end to end, all the neurones in your body would stretch for 1000 km. You would not, however, see this line, because it would be only 10 μm wide.

2 In which sense organ would you find receptor cells that detect changes in light?

3 How will the person in diagram C know that she has picked up the pencil?

4 You hear a track you like on a playlist and turn up the volume. Describe what happens in your nervous system when you do this.

The travelling, or transmission, of impulses is called **neurotransmission** and happens in **neurones** (**nerve cells**). Neurones have a cell body and long extensions to carry impulses.

There are different types of neurone. Diagram D shows a **sensory neurone**. Its function is to carry impulses from receptor cells towards the CNS. A receptor cell impulse passes into a tiny branch called a **dendrite**. It is then transmitted along the **dendron** and the **axon**. A series of **axon terminals** allow impulses to be transmitted to other neurones.

Dendrons and axons are frequently long, to allow fast neurotransmission over long distances. There is also a fatty layer surrounding these parts, called the **myelin sheath**. This electrically insulates a neurone from neighbouring neurones (e.g. in a nerve), stopping the signal losing energy. It also makes an impulse 'jump' along the cell between the gaps in the myelin, and so speeds up neurotransmission.

1 Impulses from receptor cells in the eye are transmitted by sensory neurones in the optic nerve to the brain. The brain processes these impulses and 'sees' the pencil.

2 The brain can send more impulses to tell parts of the body to do something (the response).

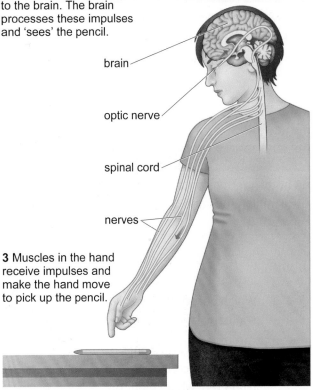

3 Muscles in the hand receive impulses and make the hand move to pick up the pencil.

C This is what happens in the nervous system when someone picks up a pencil.

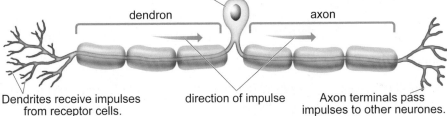
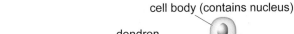

D a sensory neurone

Dendrites receive impulses from receptor cells. direction of impulse Axon terminals pass impulses to other neurones.

5 a Suggest the name of a cell that a dendrite might receive an impulse from.

b Draw a flow chart to show the route of an impulse along a sensory neurone.

6 Explain the ways in which a sensory neurone is adapted to its function.

7 Explain what is meant by a 'response to a stimulus'.

Checkpoint

How confidently can you answer the Progression questions?

Strengthen

S1 Draw a flow chart to show how information about something touching the heel of your foot gets to your brain.

Extend

E1 You pick up an ice cube. Explain how your nervous system allows you to do this.

Exam-style question

Describe how you detect the stimulus of temperature change. *(3 marks)*

SB2h The eye

Specification reference: B2.15B; B2.16B; B2.17B

Progression questions

- How do our eyes allow us to see?
- How do some eye defects change vision?
- How can some eye defects be corrected?

Did you know?

A human retina contains about 120 million rods and 6 million cones.

The eye is a sense organ that contains receptor cells found in a layer called the **retina**.

Cones are receptor cells that are sensitive to the colour of light. Some cones detect red light, while others detect green or blue. Cones generate impulses in sensory neurones, which lead into the brain through the **optic nerve**. The information from all the cones is processed into full colour vision at the back of the cerebral hemispheres.

Rods are receptor cells that detect differences in light intensity, not colour. Rods work well in very dim light whereas cones only work in bright light, which is why your colour vision is poor in dim light.

 1 Which part of the eye gives someone brown eyes?

 2 If a person's 'red' cones and 'green' cones are triggered, what colour do they see?

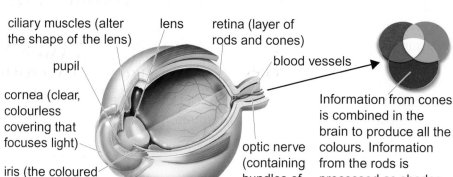

ciliary muscles (alter the shape of the lens)

lens

retina (layer of rods and cones)

pupil

blood vessels

cornea (clear, colourless covering that focuses light)

iris (the coloured part of the eye)

optic nerve (containing bundles of neurones)

Information from cones is combined in the brain to produce all the colours. Information from the rods is processed as shades of dark and light.

A the structure of the eye

The **pupil** is the dark area in the middle of the eye, and is where light enters. The amount of light entering the eye is controlled by muscles in the **iris**, which can **constrict** the pupil (decrease its diameter) or **dilate** it (make it bigger). Bright light can damage the receptor cells in the retina.

 3 a When moving from a dark room into bright light, what happens to the pupil of each eye?

 b What controls the size of the pupils?

 c Explain why this happens.

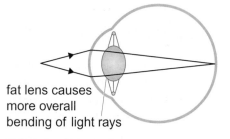

thin lens causes less overall bending of light rays

light rays refracted by cornea

light rays also refracted by lens

fat lens causes more overall bending of light rays

B focusing light in the eye

 4 Explain why the lens becomes fatter when focusing on closer objects.

Light rays entering the eye need to be focused onto a point in the retina to produce a clear image. Most focusing is done by the **cornea**, which bends (refracts) light rays to bring them together. The **lens** then fine-tunes the focusing. **Ciliary muscles** make the lens fatter to focus light from near objects and thinner to focus light from distant objects.

Eye problems

For **short-sighted** people distant objects appear blurred. This is because the rays of light from distant objects are focused in front of the retina. There are two possible reasons for this: the eyeball is too long or the cornea is too curved and bends the rays more than it should. **Long-sightedness** is caused by the opposite problems.

Contact lenses or lenses in glasses can correct these problems, or a laser can be used to cut away some of the cornea and reshape it.

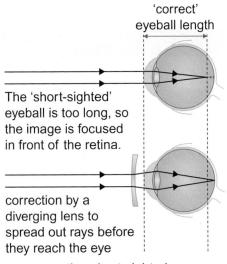

The 'short-sighted' eyeball is too long, so the image is focused in front of the retina.

correction by a diverging lens to spread out rays before they reach the eye

correcting short-sightedness with a **diverging lens**

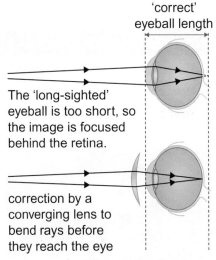

The 'long-sighted' eyeball is too short, so the image is focused behind the retina.

correction by a converging lens to bend rays before they reach the eye

correcting long-sightedness with a **converging lens**

C correcting long- and short-sightedness with lenses

 5 a Describe how someone would know if they were short-sighted.

 b Explain how short-sightedness can be corrected using glasses.

Sometimes a protein builds up inside the lens and makes it cloudy. This is a **cataract**. Full vision can be restored by replacing the clouded lens with a plastic one.

D a cloudy-looking pupil shows a cataract

People with **colour-blindness** have some cones that do not work properly and so have difficulty in seeing some colours. The most common form is red–green colour-blindness, in which the cones that detect green light are faulty, making it difficult to tell the difference between reds, greens and browns. Colour-blindness cannot be corrected.

 6 a Explain which eye the person in photo D has most difficulty seeing with.

 b Explain two important properties a material used for an artificial lens must have.

 7 Why might someone with red–green colour-blindness have difficulty in telling yellow from red?

Exam-style question

Describe how light from distant objects is focused onto the back of the retina.

(3 marks)

Checkpoint

How confidently can you answer the Progression questions?

Strengthen

S1 Draw a table to explain how different eye problems are caused and describe how they can be corrected.

Extend

E1 As people get older, the lens gets harder and less able to form a fat shape. Use diagrams to explain what problem this can cause and how it can be corrected with a lens.

SB2i Neurotransmission speeds

Specification reference: B2.13; B2.14

Progression questions

- How is a motor neurone adapted to its function?
- How do neurotransmitters allow a connection between neurones?
- How does the structure of a reflex arc allow faster reactions to stimuli?

A Lance Corporal Craig Lundberg was blinded by a grenade but gets around unaided using the BrainPort®.

The device in photo A allows blind people to see … with their tongues! The image from the camera is sent to a 'lollipop' that contains hundreds of small electrodes. Each electrode produces pulses of electricity depending on how much light is in that part of the image. By putting the lollipop on the tongue, the user can feel these pulses and build up an idea of basic shapes and movement. This allows some blind people to react and respond to visual stimuli.

When the brain coordinates a response to a stimulus, impulses are sent to **effectors** and these carry out an action. Effectors include muscles and glands (e.g. sweat glands).

1 Imagine you see a lion and run away.

    a Where are the receptor cells that receive the stimulus?

    b What effectors carry out the response?

    c Suggest another effector triggered by seeing the lion.

Different neurones

Motor neurones carry impulses to effectors. **Relay neurones** are short neurones that are found in the spinal cord, where they link motor and sensory neurones. They also make up a lot of the nerve tissue in the brain. Neither of these types of neurone has a dendron, and the dendrites are on the cell body.

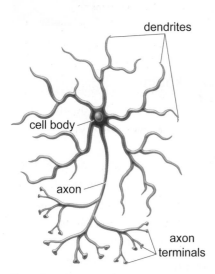

B a relay neurone

2 Do the following carry information to or away from the central nervous system?

    a motor neurones

   b sensory neurones

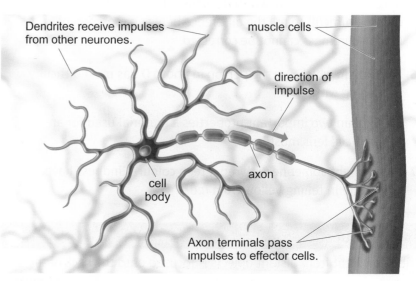

C a motor neurone

Synapses

One neurone meets another at a **synapse**, which contains a tiny gap. When an impulse reaches an axon terminal, a **neurotransmitter** substance is released into the gap. This is detected by the next neurone, which generates a new impulse. Synapses slow down neurotransmission. They are, however, useful because neurotransmitters are only released from axon terminals and so impulses only flow in one direction. They also allow many fresh impulses to be generated in many neurones connected to one neurone – the original impulse does not need to be split and lose 'strength'.

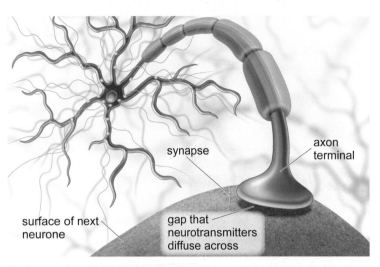

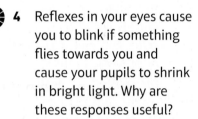

D The gap in a synapse is only about 20 nm (0.00002 mm) wide.

 3 Give two reasons why synapses are used in the nervous system.

The reflex arc

If you touch a very hot object you need to pull your finger away very quickly to stop it burning you. You don't want to have to waste time thinking about this and so a **reflex** is used. Reflex actions are responses that are automatic, extremely quick and protect the body. They use neurone pathways called **reflex arcs**, which bypass the parts of the brain involved in processing information and so are quicker than responses that need processing.

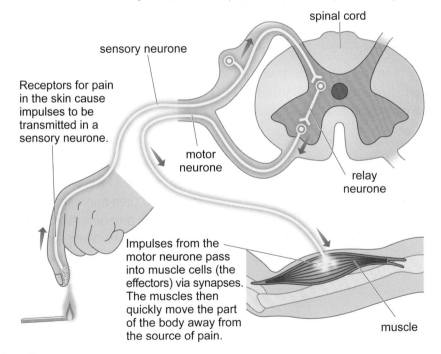

Receptors for pain in the skin cause impulses to be transmitted in a sensory neurone.

Impulses from the motor neurone pass into muscle cells (the effectors) via synapses. The muscles then quickly move the part of the body away from the source of pain.

spinal cord

sensory neurone

motor neurone

relay neurone

muscle

E a reflex arc

4 Reflexes in your eyes cause you to blink if something flies towards you and cause your pupils to shrink in bright light. Why are these responses useful?

5 Draw a table to compare and contrast reflex actions with processed responses.

6 Suggest why there are very few synapses between the receptor and effector in a reflex arc.

7 You kick a football. Describe how this response occurs.

Checkpoint

How confidently can you answer the Progression questions?

Strengthen

S1 Draw a flow chart to show how an impulse in a relay neurone causes an impulse in a motor neurone.

Extend

E1 Explain how response times are decreased in reflex actions.

Exam-style question

Describe how the arrival of an impulse at the end of one neurone can cause an impulse in a neighbouring neurone. *(3 marks)*

Reflex arc

Describe how impulses are transmitted in a reflex arc in order to reduce the chance of getting burnt when someone touches a hot object.

(6 marks)

Student answer

If you touch a hot object, receptor cells in the hand detect this. Impulses travel into a sensory neurone, and then into a relay neurone in the spinal cord. After [1] this, the impulses pass to motor neurones, which cause muscles (the effectors) to move the hand quickly out of the way [2]. This pathway is called a reflex arc, and it is shorter than if the impulses had to go to the brain to be processed. This means that the response is much quicker than usual [3]. This means that the hand is moved away from the hot object as quickly as possible, meaning that there is less chance of it being burnt [4].

[1] The answer shows good use of prepositions such as 'then' and 'after' to show order.

[2] The neurones are described in the correct order.

[3] A good explanation of why the hand moves away so quickly.

[4] The answer relates back to the question, which was about the chance of being burnt.

Verdict

This is a strong answer. The answer is arranged logically and written clearly. It shows good linking of scientific knowledge to the context of the question.

The answer first describes the reflex arc in the correct order, and uses the correct scientific names for the neurones. The answer then links the description to the speed of response, and then to the reduced chance of being burnt. This was the original point of the question.

Exam tip

When you have to write about a sequence of events, it is a good idea to write out bullet points of the facts that you will use, and then number them to show the order in which you will use the points. When you have written your answer, neatly cross out your bullet points.

Paper 1

SB3 Genetics

This image was made by splicing together a photograph of a mother (aged 52) and her son (aged 30). It is one of a number of 'genetic portraits' created by Canadian graphic designer Ulric Collette to illustrate how closely members of the same family resemble one another.

In this unit you will learn about the DNA code that produces our features and the processes that allow features to be passed on from parents to their offspring.

The learning journey

Previously you will have learnt at KS3:

- about the differences between environmental and inherited (or genetic) variation
- how two gametes fuse to produce a zygote
- how the nuclei of eukaryotic cells contain chromosomes, which contain DNA.

In this unit you will learn:

- about sexual and asexual reproduction, and the need for meiosis
- about the structure of DNA and its role in protein synthesis
- about mutations and the causes of genetic variation
- how the inheritance of some characteristics occurs in families.

SB3a Sexual and asexual reproduction

Specification reference: B3.1B; B3.2B

Progression questions

- How do asexual and sexual reproduction differ?
- When is asexual reproduction more advantageous than sexual reproduction?
- When is sexual reproduction more advantageous than asexual reproduction?

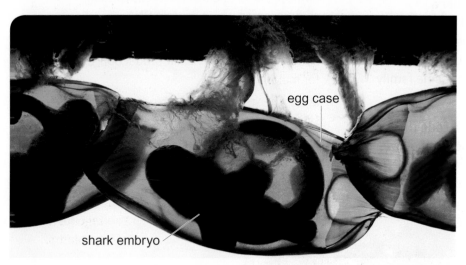

A These bamboo shark eggs were produced by a female shark kept in a zoo with only females in her tank. Genetic analysis of the embryos shows that they are clones of their mother.

Most animals and plants reproduce by **sexual reproduction**, involving **fertilisation** of a female sex cell by a male sex cell. Some organisms can reproduce without fertilisation, which is known as **asexual reproduction**. This produces **clones** (offspring that are genetically identical to the parent).

1 State, with a reason, whether humans reproduce sexually or asexually.

2 Describe what is meant by a clone.

In animals, asexual reproduction is very rare in **vertebrates** but much more common in **invertebrates** (such as insects). For example, during the summer, female aphids produce offspring from cells formed by **mitosis**.

3 Explain whether the offspring produced by asexual reproduction in aphids are male or female.

Many plants are able to reproduce asexually. Some use runners, which are special stems that grow out from the adult plant. Others produce new plants from bits of leaves or roots (see *SB2a Mitosis*). This helps them spread quickly in areas near the parent plant.

B In the summer, an adult female aphid can produce up to five young a day without the need for a mate.

Did you know?

If *Viola* plant flowers are not fertilised in spring, more flowers are produced in summer. However, summer flowers do not open and instead fertilise themselves. This makes sure that seeds are produced each year.

The runner supplies water and nutrients to the daughter plant until its leaves and roots are well developed.

Daughter plant growing where a node (joint) in a runner touches the ground. New plants grow leaves then roots.

adult plant

C New strawberry plants grow where a runner touches the ground. When the new plant has well-developed roots, the runner dies off and the plant grows on its own.

Sexual reproduction combines characteristics from both parents, and so produces offspring that are different from each other. This is an advantage if the offspring move to an area with different environmental conditions. It is also advantageous if the environment changes, for example if temperature changes or a new pest or disease comes to the area. **Variation** means that there is a greater chance that some offspring will be better suited to new conditions and so will be more likely to survive and reproduce.

Asexual reproduction is much faster than sexual reproduction because there is no need to find a mate. This is an advantage when there are a lot of resources such as food. Aphids feed off the sap of many plants. Plant growth is most rapid in summer, so asexual reproduction helps aphids make the most of all the food that becomes available.

D These trees are clones of one male aspen tree. The colony is called Pando. The clones cover an area of over 4 square kilometres, and grow from underground roots in an area where there have been frequent forest fires that destroy anything above ground.

 5 Explain how asexual reproduction is an advantage to the Pando trees.

 6 The Pando trees are dying, possibly from disease. Explain why being clones means it is more likely that they will all die than if they had grown from seeds.

 7 a Suggest how reproducing asexually could be useful to the bamboo shark.

 b Suggest why bamboo sharks usually use sexual reproduction.

Exam-style question

Describe one advantage and one disadvantage to an animal of reproducing asexually. *(2 marks)*

 4 a Explain why offspring from asexual reproduction in strawberry plants may be better adapted to conditions near the parent plant than offspring from sexual reproduction.

 b Strawberry plants also reproduce sexually, forming seeds on fruits that animals eat. The animals then usually drop the seeds far from the plant in their waste. Explain how sexual reproduction is advantageous to strawberry plants.

Checkpoint

How confidently can you answer the Progression questions?

Strengthen

S1 Construct a summary table to show the advantages and disadvantages of asexual reproduction and sexual reproduction.

Extend

E1 Late in summer, aphids produce sexual males and females. A sexual female that has mated with a male lays eggs that hatch out the next spring, producing females that reproduce asexually when they find a new food plant. Explain how both methods of reproduction are advantageous to aphids.

SB3b Meiosis

Specification reference: B3.3; B3.5

Progression questions

- What happens in meiosis?
- Why is meiosis necessary for sexual reproduction?
- What is the role of the genome in the manufacture of proteins?

You started off life smaller than this.

A A human zygote is 0.1 mm in diameter, which is even smaller than this tiny dot. (You may need a magnifying glass to see the dot!)

Did you know?

There are about 37 million million cells in an adult human.

Humans start life as a single fertilised egg cell, a **zygote**. This is formed when two **gametes** (sex cells) fuse during **fertilisation**. The zygote then forms a ball of cells using a type of cell division called **mitosis** (see *SB2a Mitosis*).

Almost all human cells carry exactly the same instructions. These instructions control each individual cell, and also shape, coordinate and control our bodies. So, it's amazing that all the instructions fit into the nucleus of the zygote (the nucleus is about 6 μm or 0.006 mm in diameter).

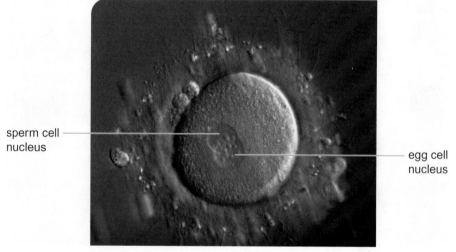

sperm cell nucleus

egg cell nucleus

B a human **sperm cell** nucleus and an **egg cell** nucleus fusing during fertilisation

The instructions for an organism are found as code in a molecule called **DNA**. The DNA of an organism is its **genome**, and most cells contain a complete copy of an organism's genome.

The human genome is found on 46 very long molecules of DNA, and each molecule is inside a **chromosome**. Along the length of a DNA molecule are sections that each contain a code for making a protein. These DNA sections are **genes**. Proteins are **polymers**, made by linking different amino acids together in a chain. The order of amino acids is controlled by a gene. Humans have about 20 000 genes.

There are 23 different chromosomes in humans and most nuclei contain two of each type. So, a human body cell contains two sets of 23 chromosomes, making 46 in all. A cell like this is **diploid** (from the Greek for 'double'). The shorthand for this is 2n. If two diploid cells joined in fertilisation, the **zygote** would have four sets of chromosomes, so gametes need to have just one set of chromosomes. They have to be **haploid**. The shorthand for a haploid cell is 1n.

1 State the names of the human gametes.

2 Name the process that turns the zygote into a ball of cells.

3 How many individual chromosomes do the following cells have?

   a human sperm cell

   b human liver cell

   c human zygote

4 How much of an organism's genome does a gamete contain?

5 Define the term 'gene'.

Gamete production

Mitosis produces diploid cells, and so a different process is used to produce gametes – **meiosis**.

Each chromosome **replicates** – makes a copy of itself. The two copies remain attached, making each chromosome look like an X. The two sets of chromosomes 'pair up', forming 23 pairs, and the pairs then separate into two new cells. Next, the two copies of a chromosome in each X-shape split into two more new cells. Meiosis therefore produces four haploid **daughter cells**, which is how gametes are produced.

Each chromosome in a pair contains different versions of the same genes. They are 'genetically different'. So, gametes are all different because they contain genetically different chromosomes. This explains why brothers and sisters often look similar but not identical (unless they are identical twins).

C Brothers and sisters have different selections of chromosomes passed down from their parents and so look different to one another.

The gamete-making cell has two sets of chromosomes. It is diploid (2n).

The chromosomes replicate (and the copies stay stuck to one another).

The cell divides into two and then into two again. Each of the final four daughter cells has a copy of one chromosome from each pair. They are haploid (1n).

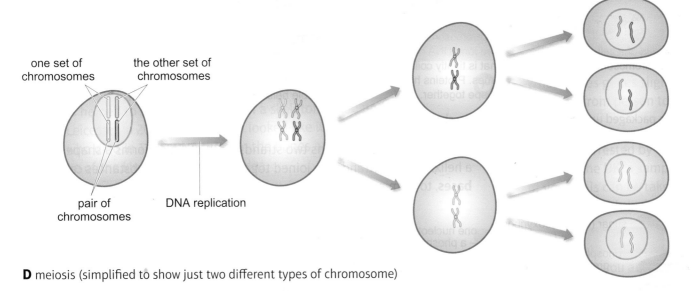

one set of chromosomes

the other set of chromosomes

pair of chromosomes

DNA replication

D meiosis (simplified to show just two different types of chromosome)

 6 Explain why chromosomes look X-shaped at the start of meiosis.

7 Name the process in which the following cells are made.

 a zygote

 b embryo cell

 c sperm cell

 8 'Triploid syndrome' is a rare human disease. What do you think the cells of someone with this disease are like?

Exam-style question

Compare and contrast mitosis and meiosis. *(3 marks)*

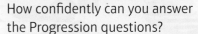

Checkpoint

How confidently can you answer the Progression questions?

Strengthen

S1 Draw a diagram to explain how a zygote's genome is created.

Extend

E1 Explain why a brother and sister do not look identical.

SB3d Protein synthesis

Specification reference: **H** B3.7B; **H** B3.8B

Progression questions

- **H** How does DNA store a code for building proteins?
- **H** What happens during the transcription stage in protein synthesis?
- **H** What happens during the translation stage in protein synthesis?

H

A Watson (on the left) and Crick with their DNA model in Cambridge

RNA polymerase

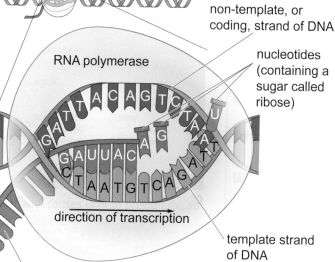

non-coding binding site | RNA polymerase binds to DNA and separates the strands.

RNA polymerase moves along the DNA and starts making mRNA when it reaches the template strand.

non-template, or coding, strand of DNA

nucleotides (containing a sugar called ribose)

RNA polymerase

direction of transcription

template strand of DNA

newly made mRNA strand

B Transcription is the first stage in protein synthesis (manufacture).

The race to work out the structure of DNA was won in 1953 by James Watson (1928–) and Francis Crick (1916–2004). They used evidence from other scientists, including:

- Erwin Chargaff (1905–2002), whose chromatography experiments showed that the amounts of A and T in an organism's DNA were the same, as were the amounts of G and C
- Rosalind Franklin (1920–1958), who took an X-ray photograph suggesting that DNA was a helix
- Jerry Donohue (1920–1985), who showed them how DNA bases could form hydrogen bonds.

1 a Give the full names of A, T, C and G.

b State the type of substance that A, T, C and G are.

2 Explain Erwin Chargaff's finding.

Transcription

Watson, Crick, Chargaff and 21 other international scientists then set about working out the **genetic code** – how the order of DNA bases caused amino acids to be joined in a certain order in a protein. They called themselves the 'RNA Tie Club' because the first stage in the process is **transcription**, in which the DNA bases are used to make a strand of **RNA (ribonucleic acid)**.

3 Where in DNA is the genetic code contained?

An enzyme called **RNA polymerase** attaches to the DNA in front of a gene in a non-coding region (so called because it does not contain code for a protein). The enzyme separates the two DNA strands.

The enzyme then moves along one DNA strand (the **template strand**) adding **complementary** RNA nucleotides. These contain the same bases as DNA except that **uracil (U)** is used instead of thymine (T). The nucleotides link to form a strand of **messenger RNA (mRNA)**.

H

Did you know?

The members of the RNA Tie Club wore ties embroidered with a helix.

Translation

The mRNA strands travel out of the nucleus through small holes in its membrane, called **nuclear pores** (shown in photo C). In the cytoplasm, the mRNA strands attach to **ribosomes**.

A ribosome moves along an mRNA strand three bases at a time. Each triplet of bases is called a **codon**. At each mRNA codon, a molecule of **transfer RNA (tRNA)** with complementary bases lines up. Each tRNA molecule carries a specific amino acid. As the ribosome moves along, it joins the amino acids from the tRNA molecules together, forming a **polypeptide**. This process is called **translation**.

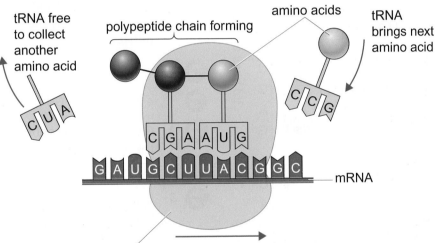

tRNA free to collect another amino acid

polypeptide chain forming

amino acids

tRNA brings next amino acid

mRNA

Ribosome moves along the mRNA in this direction, reading the code one codon at a time.

D translation

The polypeptide chain then folds up to form a protein (such as an enzyme) with a specific shape. Some proteins contain more than one polypeptide chain.

8 State how mRNA enters the cytoplasm.

9 Describe the functions of tRNA and ribosomes in translation.

10 How many codons are needed to produce a polypeptide with 23 amino acids?

Exam-style question

A gene's template strand contains the sequence ATTTCCGGTAAA. Explain how many amino acids this sequence codes for. *(2 marks)*

4 What does the genetic code actually code for?

5 What does RNA polymerase need to break in order to separate the DNA strands?

6 Describe the structure of an RNA nucleotide.

7 A template strand includes the sequence: ATTCCGGA. Write out the sequence of bases in the complementary mRNA strand.

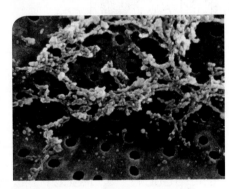

C outer surface of a nuclear membrane (×400 000)

Checkpoint

How confidently can you answer the Progression questions?

Strengthen

S1 Draw a flow chart to explain how the code in a length of DNA is used to make a protein.

Extend

E1 One DNA strand in a gene is the template strand. The other is the coding strand. Compare and contrast the coding strand with its matching mRNA strand.

SB3e Genetic variants and phenotypes

Specification reference: **H** B3.9B; **H** B3.10B

Progression questions

- **H** What is a mutation?
- **H** How can mutations alter the functions of proteins?
- **H** How can mutations alter the amount of protein that is produced?

H

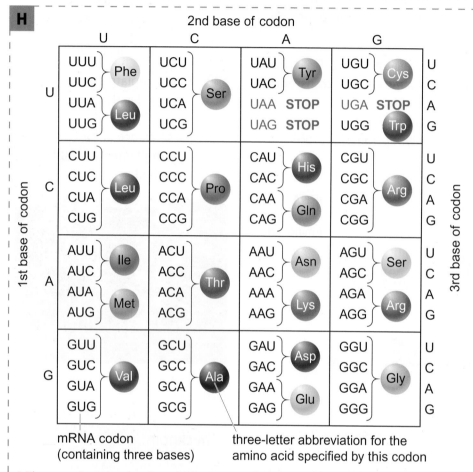

A The genetic code, showing which mRNA codons code for which amino acids. Some codons stop translation.

2nd base of codon

mRNA codon (containing three bases)

three-letter abbreviation for the amino acid specified by this codon

The RNA Tie Club discovered a lot about transcription and translation but did not crack the genetic code, which matches codons to specific amino acids. That was done by Indian-born scientist Har Gobind Khorana (1922–2011).

1 Give the three-letter symbols of the amino acids encoded by these mRNA codons.

   a UGU   **b** CCG

   c GAC   **d** GUG

2 Give the mRNA codon for the amino acid tryptophan (Trp).

3 List the amino acids in the polypeptide in diagram D on *SB3d Protein synthesis*.

4 What does the mRNA codon UAA code for?

Mutations

A change in the bases of a gene creates a genetic variant or **mutation**. It can be caused when DNA is not copied properly in cell division. Environmental factors can also cause mutations. Some mutations change an organism's **phenotype** (observable characteristics).

Mutations are the reason that genes exist in different forms, called **alleles**. One gene can have many alleles, caused by different mutations. Your characteristics are shaped by the alleles you inherit.

Haemoglobin contains four polypeptides of two kinds, α and β. The polypeptides fold and join up to form a globular (spherical) protein, which carries oxygen. Diagram B shows the mRNA made by the transcription of different alleles of the β-polypeptide gene.

Did you know?

Some people are born with extra fingers. This can be caused by a mutation in the code of a gene called *GLI3*.

mRNA from allele 1	AUG	GUG	CAU	CUG	ACU	CCU	GAG	GAG
polypeptide 1	Met	Val	His	Leu	Thr	Pro	Glu	Glu
mRNA from allele 2	AUG	GUG	CAU	CUG	ACU	CCU	GUG	GAG
polypeptide 2	Met	Val	His	Leu	Thr	Pro	Val	Glu
mRNA from allele 3	AUG	GUG	CAU	CUG	ACU	CCU	AAG	GAG
mRNA from allele 4	AUG	GUG	CAC	CUG	ACU	CCU	GAG	GAG

B mRNA from different β-polypeptide alleles

A single base change causes a different amino acid to be added.

5 Look at diagram B.

 a Explain the effect of the mutation in allele 2 on the polypeptide produced.

 b Explain why allele 3 causes problems but allele 4 does not.

Alleles 1 and 4 in diagram B result in the production of a polypeptide that folds correctly. The polypeptide from allele 2 folds incorrectly and can cause sickle cell disease (in which red blood cells become sickle shaped and stick together, resulting in episodes of extreme pain). Allele 3 also produces an incorrectly folded chain, which can make red blood cells break apart and cause shortness of breath.

The woman in photo C is explaining how babies can be tested for many **genetic disorders** (caused by mutations). Her daughter (pictured) has GA-1, a disorder caused by an enzyme that does not work but which should break down the amino acids lysine and tryptophan. In very large amounts, these damage tissues and organs. The non-functioning enzyme is due to a change in one amino acid.

 6 What is meant by the term genetic disorder?

Mutations in non-coding DNA

RNA polymerase attaches to DNA bases in front of a gene. A mutation in this non-coding region may result in RNA polymerase not binding well, reducing transcription. Such a mutation can cause β-thalassaemia, in which not enough β-polypeptide is made for haemoglobin. This causes tiredness, weakness and shortness of breath.

Other non-coding mutations can result in RNA polymerase binding better and producing more mRNA.

 7 Men with prostate cancer may have increased amounts of a protein called PSA in their blood. Suggest an explanation for this.

 8 Explain how a non-coding DNA mutation can cause the phenotype associated with β-thalassaemia.

Exam-style question

Explain why a single base change in a gene may not alter the polypeptide produced. *(1 mark)*

C Analysing genes in babies' cells can detect genetic disorders, so that treatment can start quickly.

Checkpoint

How confidently can you answer the Progression questions?

Strengthen

S1 Draw a concept map to explain how mutations can cause too much of a protein, too little, or proteins that do not function well.

Extend

E1 Part of the DNA template strand in a normal *GA-1* gene is: TAA GTG CGG GAC TAG GAA. The mutant allele has the sequence: TAA GTG CAG GAC TAG GAA. Explain why this change results in a non-functioning enzyme.

Progression questions

- Before Mendel, why did scientists struggle to understand inheritance?
- What experiments did Mendel carry out?
- Why did Mendel draw the conclusion that inheritance was due to inherited 'factors'?

 1 Describe the inherited variation in one characteristic of people close to you.

Pea shape	round	wrinkled
Pea colour	yellow	green
Pod shape	inflated	constricted
Pod colour	yellow	green
Plant height	tall	short
Flower position	terminal	axial
Flower colour	purple	white

B some of the 'non-blending' variations in characteristics in pea plants

Most offspring look like a blend of their parents. This led scientists to believe that **variations** in parents 'blended', or fused, together in their offspring. So, a mother with short fingers and a father with long fingers would have children with medium-length fingers.

Some characteristics, such as red hair, do not fit this pattern. A parent with red hair may not have red-haired children, and children with red hair may not have red-haired parents. In the 18th century scientists could not explain this.

A Red hair cannot be due to 'blending' the variations from two parents.

 2 State why scientists thought that 'blending' caused variation in inherited characteristics.

A monk called Gregor Mendel (1822–1884) started the development of modern ideas about genes, which we now know control inherited characteristics.

In his monastery garden, Mendel observed many characteristics in pea plants that, like red hair, were either present or absent but not a 'blend'. Mendel set about developing some rules for how they were inherited.

He bred (or crossed) pea plants together by using a paintbrush to move pollen (containing male gametes) from one plant to the flower of another plant. A bag was then placed over the flower on the plant and sealed. Mendel planted the seeds that formed and observed the characteristics of the offspring.

 3 Describe why pea pod colour might be described as a 'non-blending' variation.

 4 Suggest why Mendel put the pollinated flowers in bags.

Diagram C shows one of Mendel's experiments. The first generation contained only tall plants, but when he crossed two of these he got some short plants again.

After many such experiments, Mendel concluded that inherited 'factors' control the variation of characteristics. These factors exist in different versions (now called 'alleles') that do not change. A plant has two factors for each characteristic, which are either the same version or two different versions.

Plants with two factors of the same version were true-breeding. This meant that, if the plant was self-pollinated, the offspring all had the same variation as the parent.

 5 What do we call Mendel's factors today?

Mendel summarised his work in three laws of inheritance:

1 Each gamete receives only *one* factor for a characteristic.

2 The version of a factor that a gamete receives is random and does not depend on the other factors in the gamete.

3 Some versions of a factor are more powerful than others and always have an effect in the offspring.

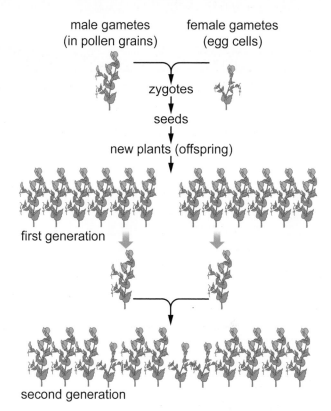

male gametes (in pollen grains)   female gametes (egg cells)

zygotes

seeds

new plants (offspring)

first generation

second generation

C one of Mendel's experimental pea plant crosses

Mendel's work was largely ignored, partly because scientists did not see how 'factors' could explain the many variations in characteristics such as human eye colour. Nor could they see how Mendel's ideas could explain Darwin's theory of evolution (see *SB4b Darwin's theory*). They argued that, if the factors could not change, then a species could not change (evolve).

Once chromosomes were discovered (in the 1880s), scientists began to see how Mendel's factors could work. His ideas started to be accepted in the 1900s and the word 'gene' was coined in 1909.

D Mendel

6 Look at diagram C.

 a Explain how Mendel could tell from his experiment that some plants in the first generation contained two versions of the factor for height, even though they were all tall.

b Which version of the pea height factor was 'more powerful'? Explain your reasoning.

 7 Explain why the importance of Mendel's work was not recognised until long after he had died.

Checkpoint

How confidently can you answer the Progression questions?

Strengthen

S1 Rewrite Mendel's laws using the terms 'gene' and 'allele'.

Extend

E1 Explain how Mendel would have worked out which version of the pea shape factor was more powerful: wrinkled peas or smooth peas.

Exam-style question

Explain why, up until the 20th century, scientists found it difficult to explain inheritance in humans. *(2 marks)*

SB3g Alleles

Specification reference: B3.12; B3.13; B3.14

Progression questions

- What is the difference between a gene and an allele?
- Why will a recessive allele not affect the phenotype of an organism that is heterozygous for that gene?
- Why are genetic diagrams useful?

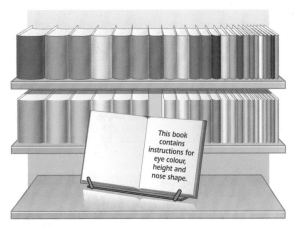

A The human genome is contained on two sets of 23 chromosomes.

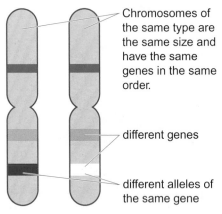

Chromosomes of the same type are the same size and have the same genes in the same order.

different genes

different alleles of the same gene

B Each gene can exist in a number of different forms called alleles.

You can think of chromosomes as a set of books. Each book (chromosome) contains a set of sentences giving instructions (genes). All of the books together contain all of the instructions needed to produce a certain organism (its genome).

1 List these terms in order of size, largest first: base, chromosome, gene, genome.

Genes for the same characteristic (e.g. eye colour) can contain slightly different instructions that create variations (e.g. brown, blue). Different forms of the same gene are called **alleles**.

Since there are two copies of every chromosome in a body cell nucleus, a body cell contains two copies of every gene. Each copy of a gene may be a different allele. There are many alleles for most of the 20 000 human genes, and the different combination of alleles in each person gives each of us slightly different characteristics (**genetic variation**).

2 Define the term 'allele'.

3 How does the idea of alleles help to explain why we all look different?

If both alleles for one gene are the same, an organism is **homozygous** for that gene. If the alleles are different, an organism is **heterozygous**. The plants on the far left and far right of diagram C are homozygous.

Gametes have only one copy of each chromosome and so only contain one copy of each gene. In diagram C, gametes from the purple and white flowers fertilise and form a **zygote** that is heterozygous. However, in the offspring plant only the allele for purple flowers has an effect. It is said to be **dominant**. The white flower allele has no effect if the purple flower allele is there. This white flower allele is **recessive**.

The flower colour alleles are both the same.
They contain the instructions for purple flowers.

gametes contain the purple allele

All the offspring have both alleles. However, all the flowers are purple.
This is because purple is the dominant allele.

gametes contain the white allele

The flower colour alleles are both the same.
They contain the instructions for white flowers.

C

A recessive characteristic is only seen if *both* alleles are recessive. This can be shown in a **genetic diagram** (diagram D).

A dominant allele is shown by a capital letter (e.g. R for purple). The recessive allele has the lower case version of the *same* letter (e.g. r for *not* purple). The letter for the dominant allele is always written before the recessive one (e.g. Rr and never rR). The alleles in an organism are its **genotype**. What the organism looks like is its **phenotype**.

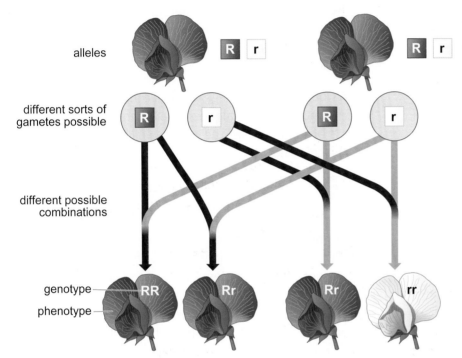

D A genetic diagram shows the possible combinations of alleles when two organisms breed. We use diagrams like this to explain the inheritance of one gene (**monohybrid inheritance**). We can also use these diagrams to predict the **ratios** of the phenotypes (purple and white phenotypes are in a 3:1 ratio in this example).

6 When will a recessive allele affect a phenotype?

7 Draw a table to show the three genotypes in diagram D and their matching phenotypes.

8 The pea plant gene for height has two alleles: T (dominant, causing tall plants) and t (recessive, causing short plants). A homozygous tall plant is crossed (bred) with a homozygous short plant. Draw a genetic diagram to show the pattern of monohybrid inheritance of the height gene.

Exam-style question

ADWH is a condition in which people have curly hair that looks like sheep's wool. It is caused by a dominant allele (D). Calculate the ratio of children with and without the condition if one parent is heterozygous and the other is homozygous for the recessive allele. *(4 marks)*

 4 How would you add to the book analogy (in diagram A) to include alleles?

 5 a Is the plant in the centre of diagram C heterozygous or homozygous for the flower colour allele?

 b Explain your reasoning.

Did you know?

The ability to taste a bitter substance called PTC (phenylthiocarbamide) is controlled by one gene. People with a dominant allele of the gene can taste PTC. People with two recessive alleles cannot taste it.

Checkpoint

How confidently can you answer the Progression questions?

Strengthen

S1 Two pea plants are bred together. One has two dominant alleles for purple flowers (R) and one has two recessive alleles for white flowers (r). Draw a diagram to explain why none of the offspring will have white flowers.

Extend

E1 The pea plant gene for seed shape has two alleles: N (causing smooth peas) and n (causing wrinkled peas). Use a genetic diagram to work out the ratio of offspring with smooth and wrinkled peas if the parent plants are both heterozygous.

SB3h Inheritance

Specification reference: B3.14; B3.15; B3.16

Progression questions

- How is the sex of offspring determined in humans?
- How do we use genetic diagrams, Punnett squares and family pedigrees to show inheritance?
- How are the probable outcomes of offspring phenotypes calculated, using information about alleles?

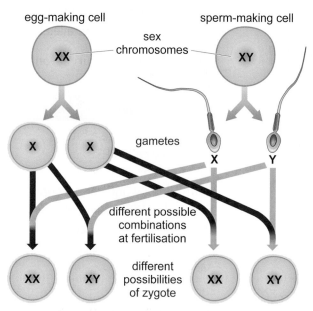

A genetic diagram for sex determination in humans

Two of your chromosomes determine what sex you are. They are your **sex chromosomes** and there are two types: X and Y. Females have two X sex chromosomes and males have one X and one Y.

A woman's gametes (egg cells) all contain an X sex chromosome but male sperm cells contain either an X or a Y. Diagram A shows the ways in which sex chromosomes can combine.

1 In humans, which gamete is responsible for determining the sex of the offspring?

Punnett squares (such as diagram B) are another way to demonstrate inheritance. The boxes show the possible genotypes. Two boxes contain XX and two contain XY, so the ratio of the outcomes is 2:2 (simplified to 1:1). This means that there should be equal numbers of offspring born with XX sex chromosomes as are born with XY.

The likelihood of an event happening is its **probability**. Punnett squares are used to work out the theoretical probability of offspring inheriting certain genotypes. Probabilities can be shown as fractions, decimals or percentages. The probability of an impossible event is 0 or 0 per cent. The probability of an event that is certain to happen is 1 or 100 per cent.

The boxes show the possible combinations in the offspring.

B Punnett square for human sex determination

Worked example

In diagram B, two out of four boxes are XX. So the probability of a child being XX is:

$$\frac{2}{4} = \frac{1}{2} = 0.5$$

Or as a percentage: 50%

Did you know?

Birds have Z and W sex chromosomes and it is the egg cell that determines the sex; the females are ZW and males are ZZ.

 2 Explain why about half the population of the UK is female.

Punnett squares are used to work out the probabilities of different phenotypes caused by alleles. Diagram C shows an example for a genetic disorder called cystic fibrosis (CF), caused by a recessive allele. People with CF have problems with their lungs and digestive systems.

People who inherit two recessive CF alleles (f) have the disorder. People who inherit at least one dominant allele (F) do not have CF. The Punnett square shows that if heterozygous parents have children, the ratio of children without CF to those with CF is predicted to be 3:1.

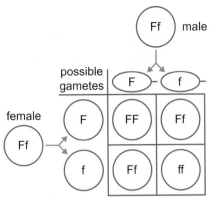

C Punnett square for parents that are heterozygous for the CF gene

 3 Draw a Punnett square for the plants shown in SB3g diagram D.

4 Look at diagram C.

a What is the ratio of the phenotypes of the offspring?

b If the parents had four children, how many are predicted to have CF?

c Calculate the probability of a child being born without CF. Show your answer as a fraction, a decimal and as a percentage, and show all your working.

A **family pedigree chart** (such as diagram D, below) shows how genotypes and their resulting phenotypes are inherited in families.

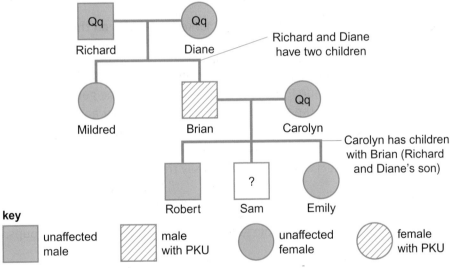

key

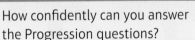

D Phenylketonuria (PKU) is a disorder caused by a recessive allele. People with PKU lack an enzyme called phenylalanine hydroxylase and can develop nerve problems unless they stick to a special diet.

5 Look at diagram D.

 a Which letter shows the recessive allele?

 b Why doesn't Carolyn have PKU?

 c What is Brian's genotype?

 d Calculate the probability that Sam has PKU. Show your working.

Exam-style question

Explain how the sperm cell is responsible for the determination of sex in humans. *(4 marks)*

SB3i Multiple and missing alleles

Specification reference: B3.17B; **H** B3.18B

Progression questions

- How are the ABO blood groups inherited?
- What is codominance?
- **H** Why do more men than women suffer from sex-linked genetic disorders?

blood type A   blood type B

blood type AB   blood type O

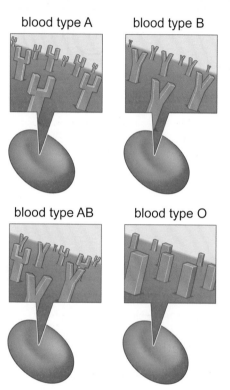

A People with different blood types have different markers on their red blood cells.

If a person loses a lot of blood, such as in an accident or operation, they must be given more blood to help them survive. The blood they are given must be of the right type otherwise the red blood cells in it will clump together, which can kill.

One way of classifying different types of blood is the **ABO blood group** system. In this system, everyone's blood is in one of four groups: A, B, AB and O. Which blood group you have is determined by whether you have certain 'marker molecules' on the outside of your red blood cells. There are three main types of these markers, which we refer to as A, B and O.

 1 Use diagram A to identify the marker molecules on the red blood cells of people with each of the four ABO blood groups.

The gene that is responsible for the markers in the ABO system has three alleles, written as I^A, I^B and I^o. Everyone has two copies of the gene, so may be homozygous for any of the three alleles or heterozygous for any two of the three alleles. I^o is recessive to both I^A and I^B. However, a person with genotype $I^A I^B$ shows the effect of both alleles and has the blood group AB. When both alleles for a gene affect the phenotype, we say they are **codominant**.

 2 Explain why ABO blood groups show codominance.

3 Write down the possible genotypes of the following blood groups.

    a AB   **b** O   **c** A   **d** B

H

Y   X

These regions of the X chromosome are missing in the Y chromosome.

B The human Y and X chromosomes.

Sex-linked genetic disorders

Chromosomes in diploid cells come in pairs. In most pairs, the chromosomes have the same genes. However, the human Y sex chromosome is missing some of the genes found on the X chromosome. This means a man (XY) will have only one allele for some genes on the X chromosome (because those genes are missing on the Y chromosome). If the allele for one of these X chromosome genes causes a genetic disorder, then a man will develop that disorder.

If a woman (XX) inherits the 'disorder' allele, she may have a 'healthy' allele on her other X chromosome. If the 'disorder' allele is recessive, she will not get the disorder. If she inherits two recessive 'disorder' alleles, she will develop the disorder. However, the probability of a woman getting the disorder is much less than that of a man developing it. Disorders that show a different pattern of inheritance in men and women are called **sex-linked genetic disorders**.

4 Norrie disease is a sex-linked disorder that causes blindness only in men. Draw the X and Y chromosomes for a man with Norrie disease. Label your drawing to show where you would expect to find the gene that causes Norrie disease. Give reasons for the position you chose.

There are many examples of sex-linked genetic disorders, including red–green colour blindness. This is where a person sees the colours red and green as being the same, which can happen if some cone cells in the retina do not work properly. About 8 per cent of men have this disorder, but only 0.5 per cent of women. The Punnett square shows how the disorder is inherited differently in men and women. In this example, the mother is a **carrier** for the disorder, which means she has one faulty allele and one normal allele.

 5 a In the Punnett square in diagram C, explain whether either of the parents have the red–green colour blindness phenotype.

 b Use the Punnett square to calculate the probability that a female child or a male child of these parents will develop red–green colour blindness.

 c Explain the difference in inheritance of red–green colour blindness by males and females.

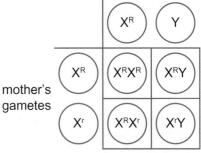

father's gametes

mother's gametes

C In this Punnett square, the allele for colour blindness on the X chromosome is written as X^r and the allele for normal colour vision is written as X^R. The Y chromosome does not have an allele for this gene.

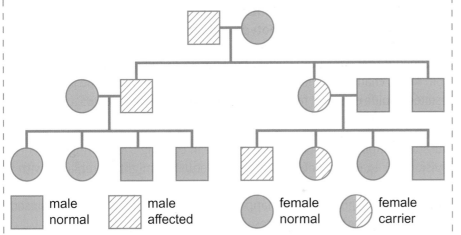

| male normal | male affected | female normal | female carrier |

D This family pedigree shows the inheritance of haemophilia in one family. Haemophilia causes bleeding that is difficult to stop.

 6 a Explain how diagram D provides evidence that haemophilia is a sex-linked disorder.

 b Explain why a female carrier of the haemophilia allele does not suffer from the disorder.

Checkpoint

How confidently can you answer the Progression questions?

Strengthen

S1 Using a Punnett square, explain why a mother with blood group A and a father with blood group B could have a child with any of the four blood groups.

Extend

E1 **H** About 1 in about 3600 men suffer from a muscle-wasting disease called Duchenne muscular dystrophy, but very few women do. Suggest where the gene for the disease is found in the genome and draw diagrams to show how it is inherited in males and females.

Exam-style question

Explain how ABO blood group alleles illustrate dominance, recessiveness and codominance. *(3 marks)*

Progression questions

- Why is it difficult to identify how most inherited characteristics are controlled?
- What is a mutation?
- How can mutations cause variation?

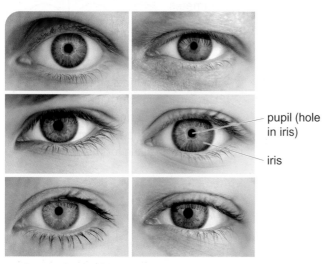

A Human eye colour is caused by the amount of melanin (a dark-coloured protein) and how light is scattered by the iris.

Most human characteristics are controlled by many genes, not just one. For example, several genes affect eye colour in humans. However, most **variation** in eye colour is caused by the OCA2 gene, which controls the amount of melanin produced. Melanin is a protein that makes hair, eyes and skin darker in colour. A blue iris contains little melanin, while a brown iris contains a lot of melanin.

 1 Suggest two alleles for the OCA2 gene, and explain your answer.

 2 It is sometimes said that brown eyes are dominant to blue eyes. Use this information, and your answer to question 1, to draw a Punnett square that describes the inheritance of eye colour from heterozygous brown-eyed parents. Describe the outcomes in terms of genotype and phenotype.

Did you know?

A baby has about 70 mutations that neither of its parents have. These mutations occur during gamete production and during growth.

It is thought that early humans had brown eyes. As our ancestors moved north from Africa, changes in the alleles for melanin production in some people resulted in blue eyes and fair skin. A change in a gene that creates a new allele is called a **mutation**. Mutations often occur during cell division.

Mutations happen when there is a mistake in copying DNA during cell division. For example, one base in a DNA sequence might be replaced with another, rather like typing the wrong letter in a word. This can happen naturally, but is more likely to happen if there is damage to the DNA caused by radiation or certain substances.

3 Which kind of cell division could produce a mutation in:

 a a gamete

 b a body cell?

Sometimes a mutation produces an allele that causes a big change in the protein that is produced. This will affect how the body works. However, mutations can occur in different parts of a gene and so may only have a small effect on the protein that is produced. Many other mutations will not change the protein at all and so have no effect on the phenotype.

 4 Explain why the risk of developing skin cancer can increase with the amount of sunlight your skin receives.

 5 Explain why a mutation does not always produce a change in a characteristic.

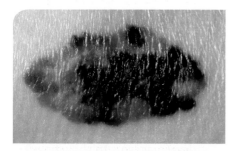

B Sunlight contains ultraviolet radiation that can cause mutations in skin cells, which can result in skin cancer.

The Human Genome Project

In 2003, the first complete human genome was decoded. This was the result of the **Human Genome Project** and involved scientists in many different countries. The project produced a map of 3.3 billion complementary base pairs in one set of 46 human chromosomes. Further work has found many sections of DNA that are the genes. Now other human genomes have also been mapped. This has shown that there are variations between people, but that over 99 per cent of the DNA bases in different people are the same.

Mapping a person's genome can indicate their risk of developing diseases that are caused by different alleles of genes. It can also help identify which medicines might be best to treat a person's illness, because the alleles we have can affect how medicines work in the body.

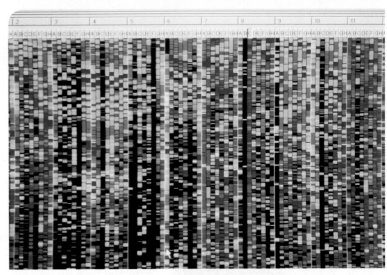

C A map of part of one human genome. Each coloured band represents a different base in the DNA sequence. Some of the bases in this part of the genome will be different in different people.

Clopidogrel is a drug used to prevent blood clots in people at risk of heart attack or stroke. Some people have particular alleles that mean the drug does not protect them.

Drug response	Status
Clopidogrel effectiveness	reduced
Simvastatin-induced myopathy	typical risk
Warfarin sensitivity	increased

The drug simvastatin is used to reduce high levels of cholesterol in the blood. Some people suffer a side effect of myopathy (weak muscles) if they take the drug.

Warfarin is given to people to reduce the risk of blood clots. Some people are more sensitive to the drug than others, due to their alleles, and so need to take a smaller dose than others, otherwise it can cause dangerous bleeding.

D Genome analysis for one person showing how they might respond to some drugs as a result of the alleles they have for particular genes.

 7 Give two ways in which information about a person's genome could be useful in medicine.

 8 a Suggest how a doctor might use the information in table D to identify which drugs to use with that particular person.

 b Explain why the information in table D can be different for different people.

Exam-style question

Explain how mutation can lead to variation in the phenotype.   *(3 marks)*

 6 Look at photo C. Explain why some colour bands would be different in this part of the genome from another person.

Checkpoint

How confidently can you answer the Progression questions?

Strengthen

S1 Different people have different eye colours. Explain why there is variation in human eye colour.

Extend

E1 The RHO gene controls the production of a protein in the retina of the eye that helps with vision in low light. The gene is 6705 bases long, and over 150 different variations in some of the bases have been found. Some variations have no obvious effect, but some lead to night blindness by the age of 40. Explain this variation in vision in different people.

SB3k Variation

Specification reference: B3.20

Progression questions

- How is genetic variation caused?
- How can the environment affect characteristics?
- What are discontinuous and continuous variation?

A These plants are all the same species, *Begonia rex*. X-ray radiation was used to produce mutations that led to many new varieties of the species with different shapes and colours of leaves.

Some of the variation between individuals of the same species is the result of variation in their genes. **Genetic variation** is caused by the different alleles inherited during sexual reproduction. Different alleles are produced by mutations, some of which cause changes in the phenotype. However, many characteristics also show **environmental variation**, because they are affected by their surroundings. For example, how well a plant grows is affected by how much light, water and nutrients it gets.

 1 a i Look at photo A. Identify one characteristic in *Begonia rex* plants that is caused by genetic variation.

 ii Explain your answer.

 b i Suggest one characteristic in these plants that shows variation caused by the environment.

 ii Explain your answer.

In a few cases, a characteristic shows variation that is caused only by the environment, such as the loss of a limb in an accident. Characteristics that are changed by the environment during the life of the individual are called **acquired characteristics**.

2 For each of these human characteristics, state whether they are controlled only by genes, only by the environment, or by both genes and environment. Explain each of your answers.

 a length of hair

 b blood group

 c height

 3 Explain why the shapes of the bushes in photo B are acquired characteristics.

B These bushes have acquired different shapes due to pruning by the gardener.

Variation can be grouped into two types:

- **discontinuous variation** is where the data can only take a limited set of values
- **continuous variation** is where the data can be any value in a range.

Charts C and D show how to display these different types of variation.

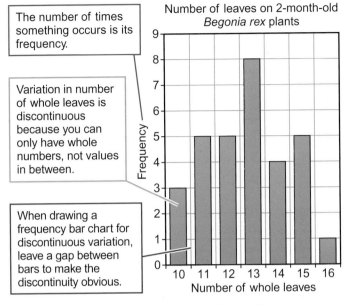

The number of times something occurs is its frequency.

Variation in number of whole leaves is discontinuous because you can only have whole numbers, not values in between.

When drawing a frequency bar chart for discontinuous variation, leave a gap between bars to make the discontinuity obvious.

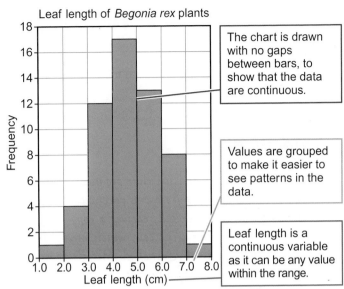

The chart is drawn with no gaps between bars, to show that the data are continuous.

Values are grouped to make it easier to see patterns in the data.

Leaf length is a continuous variable as it can be any value within the range.

C This bar chart shows variation in numbers of leaves on a plant. Since the variable on the y-axis is frequency, it is also called a frequency diagram.

D This bar chart is also a frequency diagram, and shows variation in leaf length from several *Begonia rex* plants.

4 Look at charts C and D. Which type of chart would you use for these human characteristics? Explain your answers.

 a length of hand

 b presence or absence of freckles

Continuous data for variation often forms a bell-shaped curve, known as a **normal distribution** (see graph E). It is called this because it is what is expected for a large amount of data for a characteristic where:

* the most common value is the middle value in the whole **range**
* the further a value is from the median, the fewer individuals have that value.

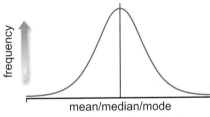

E In a normal distribution curve, the **mean** value is the same as the **mode** (most common value) and the **median** (the middle value).

5 Look at chart D.

 a What is the range of leaf lengths shown in the chart?

b What is the modal class for leaf length shown in the chart?

c Does leaf length in *Begonia* plants show a normal distribution? Explain your answer.

Exam-style question

A group of humans will show variation in height. Describe how this variation is caused.

(2 marks)

Mitosis and meiosis

Compare the roles of mitosis and meiosis in reproduction.

(6 marks)

Student answer

Mitosis is the cell division that produces two diploid daughter cells from one parent cell. During mitosis each daughter cell receives a copy of every chromosome in the parent cell, which means that [1] they are genetically identical. Mitosis is used in asexual reproduction and produces offspring that are genetically identical to the one parent [2].

In meiosis, four daughter cells are produced from one parent cell. Each daughter cell is haploid [3], and has copies of only half the chromosomes of the diploid parent cell. This produces gametes that are genetically different to each other [4]. Meiosis occurs before sexual reproduction, in which two gametes fuse to form a fertilised egg cell that is diploid. The variation in the gametes means that the offspring differ genetically from each other and from the two parents [5].

[1] The student uses a linking phrase ('which means that') to identify cause and effect. This helps to structure the answer and provide links to reasoning.

[2] This part of the answer clearly links mitosis to reproduction – which was one aim of the question.

[3] The answer correctly uses (and correctly spells) scientific terms, such as haploid.

[4] This part of the answer clearly links meiosis to cell division – which was the other aim of the question.

[5] The answer is logically divided into two paragraphs – one about mitosis and one about meiosis. Each of the paragraphs is structured in the same order. This makes the differences between mitosis and meiosis very clear.

Verdict

This is a strong answer. It contains detail about both types of cell division, and uses appropriate scientific terminology.

The answer also clearly links the description of each type of cell division with what this means for reproduction. The answer is well organised and makes it easy to compare the differences between each process.

Exam tip

The answer to a 'compare' question should describe the similarities *or* differences between **all** the subjects of the question (in this case the answer includes differences between *both* types of cell division). A 'compare' question does not need a conclusion.

Paper 1

SB4 Natural Selection and Genetic Modification

The green glow from this cat is not the result of some terrible nuclear accident! The cat has been genetically modified by scientists who are trying to stop cats getting a form of AIDS (from a virus like HIV). The cats were modified to contain a gene that helps prevent certain monkeys getting a form of AIDS. This gene was attached to another gene that makes the cats glow green under ultraviolet light. If the cat glows green, the scientists know that the modification has worked and that the cat also contains the anti-AIDS gene. In this unit you will find out more about how organisms are changed genetically by natural selection and by humans.

The learning journey

Previously you will have learnt at KS3:

- that organisms change over time (evolution)
- that Charles Darwin came up with a theory to explain this
- about how DNA contains instructions for the characteristics of organisms.

In this unit you will learn:

- about the development of the theory of evolution by natural selection
- how different methods, including genetic analysis, are being used to investigate evolution
- how organisms are classified
- how selective breeding and genetic engineering are carried out, and their benefits and drawbacks
- why tissue culture, GMOs, fertilisers and biological control are used in agriculture.

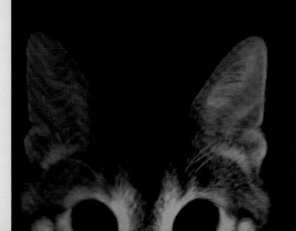

Specification reference: B4.4; B4.5

Progression questions

- What is evolution?
- How do fossils provide evidence for human evolution?
- How do stone tools provide evidence for human evolution?

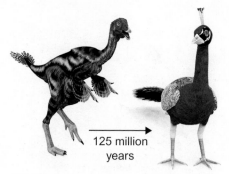

125 million years

Caudipteryx species dinosaur

Indian peafowl (*Pavo cristatus*)

A Scientists think that birds evolved from dinosaurs.

Until the 17th century, most Europeans did not think humans were animals. Ideas started to change when Carl Linnaeus (1707–1778) published his system of classification and suggested that humans were related to apes and monkeys. His **binomial system**, using two Latin words for naming **species**, is still used today.

 1 Give the binomial name for a peafowl.

Later, James Burnett, Lord Monboddo (1714–1799), proposed that humans had evolved from monkeys or apes. However, most people thought he was mad, and he had an obsession for searching for humans with tails! Now, fossil evidence suggests he was on the right track.

Fossil evidence

Evolution is a gradual change in the characteristics of a species over time. Scientists use fossils to find out about human evolution. They work out how old the fossils are and put them in age order. The fossils, though, do not show smooth changes over time because some have not been discovered.

 2 Define the term 'evolution'.

 3 Describe two differences between Ardi and a modern human.

In 1992 scientists discovered some 4.4-million-year-old fossilised bones from a female of an extinct human-like species. More fossils of this species were found and named *Ardipithecus ramidus*. The most complete set of these fossils is nicknamed **Ardi**.

Ardi was about 1.2 m tall and 50 kg. Her leg bones show that she may have been able to walk upright. She had very long arms, though, and very long big toes that stuck out to the sides of her feet and would have allowed her to climb trees.

Australopithecus afarensis (nickname **Lucy**) was discovered in 1974. She lived 3.2 million years ago and was about 1.07 m tall. She could probably walk upright, but although her toe bones were arranged in the same way as those of modern humans, they were much more curved.

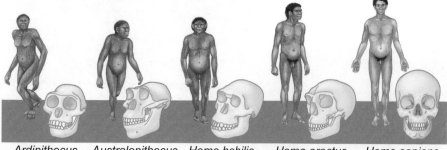

| *Ardipithecus ramidus* ('Ardi') Skull volume: 350 cm³ | *Australopithecus afarensis* ('Lucy') Skull volume: 400 cm³ | *Homo habilis* Skull volume: 500–600 cm³ | *Homo erectus* Skull volume: 850 cm³ | *Homo sapiens* Skull volume: 1450 cm³ |

B These species are arranged in age order (oldest on the left). Although there are trends, gaps in fossil evidence mean that scientists cannot be certain that these species evolved into each other.

 4 Describe one difference between Lucy and Ardi.

 5 Identify two trends in the evolution of humans.

In the 1960s, Mary Leakey (1913–1996) and Louis Leakey (1903–1972) found a more recent human-like species. They decided it was closely related to modern humans (*Homo sapiens*) and so gave it the same first word for its binomial name. It is called *Homo habilis*, which translates as 'handy man'.

Homo habilis fossils are 2.4–1.4 million years old. The animals were quite short with long arms but walked upright.

Homo erectus was discovered in Asia in the late 19th century and so many scientists thought that modern humans evolved in Asia. However, an almost complete 1.6-million-year-old skeleton was found by Richard Leakey (1944–) in 1984 in Kenya, providing evidence that humans evolved in Africa. This species was tall (1.79 m) and strongly built.

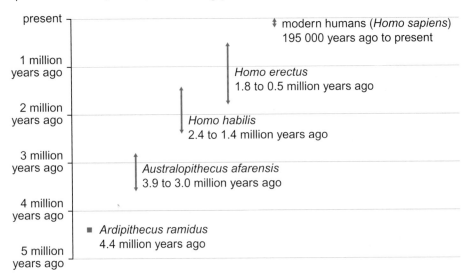

C ages of human-like species fossils

Stone tools

The earliest evidence of human-like animals using stone tools dates to about 3.3 million years ago. Scientists can work out the ages of different layers of rock. They then assume that a stone tool is about the same age as that layer of rock.

The oldest stone tools are very simple, but would have helped with skinning an animal or cutting up meat. Tools found in more recent rocks are more sophisticated.

 6 **a** Describe how stone tools developed over time.

 b Suggest a conclusion we could draw from this evidence.

 7 Some scientists suggest that there is direct evolution from *Homo habilis* to *Homo erectus* to *Homo sapiens*. Discuss arguments for and against this idea.

Exam-style question

Explain why scientists cannot be sure that human-like animals, such as Ardi, evolved into modern humans. *(2 marks)*

Did you know?

In 2015, scientists in Ethiopia found a species of human-like animal with a thick jaw and small front teeth. They think it lived alongside Lucy and have called it *Australopithecus deyiremeda*.

D The stone tool in the upper photo is about 2 million years old. The lower one is about 13 000 years old.

Checkpoint

How confidently can you answer the Progression questions?

Strengthen

S1 Describe how scientists try to show human evolution by placing fossils in order.

Extend

E1 One explanation for the appearance of more sophisticated tools is that larger or more complex brains were evolving. Suggest another hypothesis to explain this.

Progression questions

- What is natural selection?
- How does natural selection lead to evolution?
- How is Darwin's theory supported by evidence?

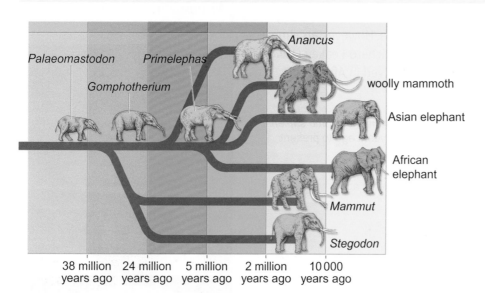

A An idea about elephant evolution, based on fossil evidence. Evolution rarely happens in one neat line – there are usually many branches.

During the 18th century, people started to accept that organisms slowly evolve into others. Two scientists, Charles Darwin (1809–1882) and Alfred Russel Wallace (1823–1913) came up with essentially the same idea about how this happened. The first book about this idea was written by Darwin and published in 1859.

1 Describe how one characteristic of African elephants has evolved.

We can think about Darwin's idea as a series of stages, as follows.

- **Genetic variation**: the characteristics of individuals vary (due to differences in genes).
- Environmental change: conditions in an area change. For example, the lack of a resource (such as food) causes more **competition** between organisms.
- **Natural selection**: by chance, the variations of some individuals make them better at coping with the change than others, and more likely to survive (also called 'survival of the fittest').
- Inheritance: the survivors breed and pass on their variations to their offspring. So the next generation contains more individuals with the 'better-adapted variations'.
- Evolution: if the environmental conditions remain changed, natural selection occurs over and over again, and a new species evolves with all the individuals having the 'better-adapted variations'.

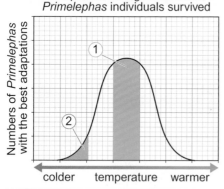

1. Most *Primelephas* were best adapted to medium temperatures. However, in conditions of medium temperature and ample resources, *all* the animals in the population could survive and reproduce.	2. Some *Primelephas* were better adapted to colder temperatures because they had more hair. When the area in which they were living became colder, more of these animals survived and reproduced.

B When conditions change, some individuals are better adapted to cope than others.

2 Suggest a characteristic that shows genetic variation in elephants.

Woolly mammoths and elephants evolved from the same animal; they share a common **ancestor**. Scientists think that an area in which this ancestor lived started getting colder. Due to genetic variation, some animals by chance had hairier skin. They were more likely to survive the cold than less hairy animals, especially when food was scarce. More of these individuals survived and bred. Over time the animals became hairier and hairier, forming a new species.

3 Suggest why elephants with longer trunks survive better than others when there is not much food.

 4 Give the name of the common ancestor of elephants and woolly mammoths.

5 Large ears help animals to cool down. Suggest an explanation for how African elephants evolved large ears.

Faster evolution

In the 1940s and 1950s, a substance called warfarin was used to poison rats. When it was first used, most rats died, but within 10 years most rats were **resistant** to (not affected by) warfarin. Due to genetic variation there had always been some rats that were resistant. As the poison killed the non-resistant rats, the only ones left to breed were resistant.

The same thing has happened with bacteria and **antibiotics** (drugs that kill bacteria). In a population of bacteria, some bacteria are more resistant than others and take longer to be killed. People who take an antibiotic to treat an infection often stop taking it too early, because they feel better. This leaves resistant bacteria still alive. They reproduce and spread, causing infections that cannot be treated with the antibiotic because all the bacteria are now resistant.

This problem of resistance in bacteria was not present when antibiotics were first used.

 7 Explain how antibiotic resistance in bacteria provides evidence to support Darwin's theory.

Bacteria in a population show variation in the amount of resistance to an antibiotic.

With time, the antibiotic kills more and more of the bacteria. The most resistant bacteria take the longest to die.

The resistant bacteria survive and reproduce. The new population of bacteria are all now resistant to the antibiotic.

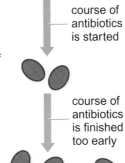

course of antibiotics is started

course of antibiotics is finished too early

key

low resistance   high resistance

D Stopping an antibiotic early can cause resistance to develop in a species of bacterium.

 6 a What environmental change in the 1940s and 1950s caused rats to evolve?

 b What allowed the rats to evolve?

Checkpoint

How confidently can you answer the Progression questions?

Strengthen

S1 List the stages of evolution and use each stage to show how bacteria develop antibiotic resistance.

Extend

E1 Ground finches have large, powerful beaks to crush seeds. A closely related species has a narrow beak for probing in small holes for insect larvae. Suggest how this species could have evolved from the seed-eating species.

Exam-style question

Wolbachia bacteria kill male blue moon butterflies. The bacteria arrived in the Samoan Islands in the late 1990s and, by 2001, only 1 per cent of the butterflies were male. Today, 50 per cent of the butterflies are male, as expected. Explain what has happened. *(3 marks)*

SB4c Development of Darwin's theory

Specification reference: B4.1B; B4.6B

Progression questions

- How did Darwin and Wallace come up with the idea of natural selection?
- What impact has the idea of evolution by natural selection had on modern biology?
- How does evidence of changes in vertebrate limbs over time support evolution by natural selection?

A Santiago Island mockingbird

B Española Island mockingbird

At the beginning of the 19th century, many scientists believed that a 'god' had created all the Earth's species, and that their characteristics could not change (evolve).

In 1835, Charles Darwin visited the Galapagos Islands, where he noticed differences between mockingbirds on different islands. He wondered whether a species could change form if it moved into a different area. He collected birds from many islands and brought them back to London, where he continued to think about evolution.

 1 Describe two differences between the mockingbirds in photos A and B.

In 1838, Darwin read an essay by Thomas Malthus (1766–1834), which argued that, if people had too many children, there would not be enough food. In the resulting struggle for survival, some children would die. This gave Darwin the idea that organisms normally produced more offspring than could survive. Only those individuals best suited to the surroundings would survive and reproduce to pass on their characteristics.

Darwin knew that this would be a controversial idea and so spent 20 years slowly piecing together evidence and writing a book. This included a careful analysis of the bird specimens he had collected.

 2 What did Darwin call the process by which better-adapted organisms are more likely to survive?

 3 Why was Darwin's idea controversial?

C Wallace noticed that tiger beetles in Indonesia had different colours depending on where they lived, to camouflage them. This puzzled him at first but his later idea about natural selection allowed him to explain it.

In 1858, Darwin received a letter from Alfred Russel Wallace, who was studying organisms in Indonesia. Wallace had also read Malthus' essay and had come to the same conclusion. Darwin quickly wrote a summary of his ideas, which was published along with Wallace's letter in a scientific paper.

 4 Suggest how Wallace might have explained the evolution of the different colours of tiger beetles, as shown in photo C.

The following year, Darwin finished his book *On the Origin of Species.* It was a bestseller, which is mainly why Darwin's name is better known than Wallace's. The theory was slow to be accepted because it challenged the belief in creation by a god. Furthermore, Darwin could not explain how variation occurred, and the evolution of characteristics in fossils was not gradual (there were sudden jumps due to a lack of fossils discovered).

The pentadactyl limb

Near the end of his book, Darwin considers the fact that vertebrates have limbs with five fingers – a **pentadactyl limb**. He asks 'Why should similar bones have been created in the formation of a wing and a leg ... as they are for totally different purposes?' The point he is making is that a human would design structures differently for flying, swimming and walking. The limb bone similarities suggest evolution from a common ancestor and not that the bones were designed for specific purposes independently of one another.

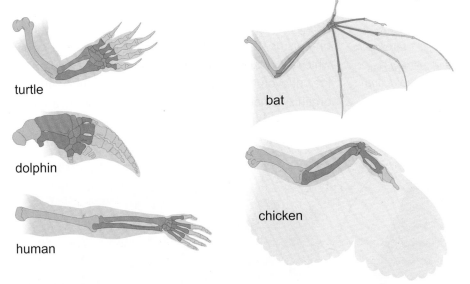

turtle

dolphin

human

bat

chicken

D Over millions of years, natural selection has caused bones in the pentadactyl limb to fuse together and change shape, but the basic similarities are there.

Darwin's book eventually changed biology. After Darwin, biologists started to think about how and why organisms changed (rather than just observing them), and especially how inherited variation occurred. This led to the development of modern 'molecular biology' and genetics.

7 Explain why Darwin's ideas led to the development of genetics.

8 How does Darwin and Wallace's theory explain the existence of very similar structures in very different organisms?

Exam-style question

Describe how the evidence in diagram D supports the idea that all these organisms evolved from a common ancestor. *(3 marks)*

5 Explain the long gap between Darwin's trip to the Galapagos and the publication of his book.

Did you know?

Scientists think that 99 per cent of all species that have ever lived are now extinct.

6 Compare and contrast the limb bones in diagram D.

Checkpoint

How confidently can you answer the Progression questions?

Strengthen

S1 Draw a time line to explain how Darwin thought of and developed his idea.

Extend

E1 Explain how Darwin and Wallace developed Malthus' idea about the struggle for survival into the theory of natural selection.

SB4d Classification

Specification reference: B4.7

Progression questions

- How are organisms classified as five kingdoms?
- How has genetic analysis changed our understanding of evolution?
- How are organisms classified as three domains?

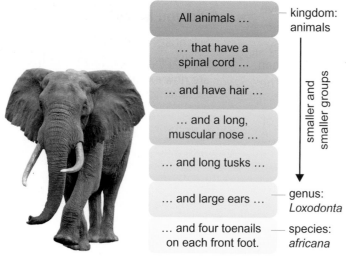

All animals …	kingdom: animals
… that have a spinal cord …	
… and have hair …	smaller and smaller groups
… and a long, muscular nose …	
… and long tusks …	
… and large ears …	genus: *Loxodonta*
… and four toenails on each front foot.	species: *africana*

A Classification today is still based on Linnaeus' original system.

In 1735, Carl Linnaeus published his **classification** system, dividing organisms into groups based on what they looked like. His largest groups (**kingdoms**) were plants and animals, which were divided into ever smaller and smaller groups. The characteristics of the organisms in a group got more and more similar as the groups got smaller and smaller. The last group contained one type of organism. Linnaeus used the two last groups (**genus** and **species**) to give each organism its binomial name.

Using characteristics for classification causes problems for organisms that have evolved similar characteristics but which are not closely related. Once scientists accepted the idea of evolution, they started to work out how different organisms had evolved and to alter the classification system so that the smaller groups contained organisms that had all evolved from recent common ancestors.

1 a What kingdom do African elephants belong to?

b What genus do they belong to?

c What is their scientific name?

2 Suggest why some people used to think that bats should be classified in the same group as birds.

3 a Look at photo B. Suggest why the fossa was originally grouped with cats.

b Suggest a piece of evidence that might have been used to show that the fossa should not be in this group.

B The fossa (from Madagascar) was classified with cats in the past. We now know that it evolved from a different ancestor to cats.

Together with an understanding of evolution, scientists since Linnaeus' time have been able to look at organisms in much greater detail, which has also helped to improve classification. Linnaeus had two kingdoms, but today we often use five kingdoms based on what the cells of organisms look like.

4 Green seaweeds do not have cellulose cell walls but do photosynthesise.

 a Suggest why seaweeds were once thought to be plants.

 b Give one reason why seaweeds are no longer classified as plants.

In the 1970s, scientists started to find examples of a new group of single-celled organisms. The cells had no nuclei, and so scientists put them into the prokaryote kingdom, as a group called Archaea.

However, scientists later found that certain Archaea genes were more similar to the genes of plants and animals than those of prokaryotes. The development of genetic analysis also showed that all organisms *apart from* prokaryotes have unused sections of DNA in their genes. Most of a gene is used to make a protein, but these 'unused' sections do not help with this. Archaea were found to have genes containing unused sections. This led Carl Woese (1928–2012) to propose that all organisms should be divided into three **domains**:

- Archaea (cells with no nucleus, genes contain unused sections of DNA)
- Bacteria (cells with no nucleus, no unused sections in genes)
- Eukarya (cells with a nucleus, unused sections in genes).

DNA changes slowly over time and so, by looking at these changes, scientists can work out how closely related two organisms are. The more DNA two organisms have in common, the more recently they evolved from a common ancestor and the more closely related they are. As DNA analysis gets faster and more precise, our classification system is updated to reflect new discoveries.

Kingdom	Main charateristics
animals	multicellular (with cells arranged as tissues and organs), cells have nuclei, no cell walls
plants	multicellular (with cells arranged as tissues and organs), have chloroplasts for photosynthesis, cells have nuclei, cellulose cell walls
fungi	multicellular (apart from yeasts), live in or on the dead matter on which they feed, cells have nuclei, cell walls contain chitin (not cellulose)
protists	mostly unicellular (a few are multicellular), cells have nuclei, some have cell walls (made of different substances but not chitin)
prokaryotes	unicellular, cells do not have nuclei, flexible cell walls

C the five-kingdom system of classification

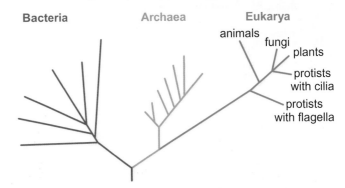

D the three-domain system of classification

 5 Which is more recent, the common ancestor of Eukarya and Archaea, or of Eukarya and Bacteria? Explain your answer with reference to diagram D.

 6 What is meant by 'genetic analysis'?

7 Explain why all members of the animal kingdom are in the Eukarya domain.

8 Explain why Archaea are no longer grouped with other bacteria.

Checkpoint

How confidently can you answer the Progression questions?

Strengthen

S1 State two things that scientists examine in order to put organisms into groups.

Extend

E1 Explain why Archaea were placed in their own domain only after genetic analysis became available.

Exam-style question

Explain how advances in technology have changed the classification of organisms.
(2 marks)

SB4e Breeds and varieties

Specification reference: B4.8; B4.10

Progression questions

- What are the ways in which we create new breeds and varieties?
- How is selective breeding carried out?
- Why do we genetically engineer organisms?

A a mouflon – the ancestor of domestic sheep

By chance, some individuals inherit characteristics that allow them to survive better than others in a certain area. This is natural selection. **Artificial selection** is when humans choose certain organisms because they have useful characteristics, such as sheep with thick wool.

About 8000 years ago, people started to look for wild sheep that were naturally hairier than others and to breed them together. They then selected the most hairy offspring and used them to breed. By repeating this over and over again, they eventually ended up with woolly sheep. Breeding organisms in this way is called **selective breeding**. It is still done today to produce new **breeds** of animal species and new **varieties** of plant species.

1 Suggest two characteristics a cattle breeder might select for.

2 Explain how a goat that produces more milk could be selectively bred.

One of the first plants to be selectively bred was wheat, about 12 000 years ago. Wild wheat plants produce few grains (seeds), and the grains fall off the plant when ripe. So early farmers selectively bred new varieties of wheat that had more grains that stayed on the plants, making them easier to harvest.

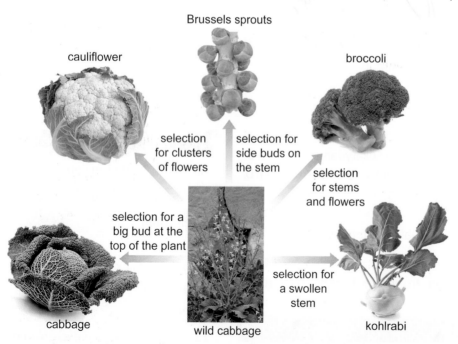

cauliflower

Brussels sprouts

broccoli

selection for clusters of flowers

selection for side buds on the stem

selection for stems and flowers

selection for a big bud at the top of the plant

selection for a swollen stem

cabbage

wild cabbage

kohlrabi

B Selective breeding of wild cabbage has produced many vegetables – all varieties of the same species.

Plants and animals are often selectively bred for:

- **disease resistance** (how well they cope with diseases)
- **yield** (how much useful product they make)
- coping with certain environmental conditions
- fast growth
- flavour.

3 Suggest two characteristics a wheat breeder might select for.

4 Explain how kohlrabi has been selectively bred.

Genetic engineering

Genetic engineering involves changing the DNA of one organism (its **genome**), often by inserting **genes** from another. This creates **genetically modified organisms** (**GMOs**). The process is much faster than artificial selection but much more expensive.

Golden Rice is a GMO with two genes inserted into its genome, one from a daffodil and one from a bacterium. They allow the rice to produce beta-carotene in its grains. Humans need beta-carotene to make vitamin A, a lack of which can cause blindness. The scientists who created Golden Rice hope that it can be grown by farmers in poorer parts of the world where vitamin A deficiency is a problem.

C Beta-carotene makes Golden Rice yellower than normal rice.

Some GMOs are resistant to disease-causing organisms, and others grow larger and faster than normal. Scientists are developing GM goats and sheep to produce proteins in their milk that can treat human diseases. GM pigs are being developed with human-like organs to use in organ transplants.

 5 Explain how Golden Rice could help reduce blindness in poorer parts of the world.

AquAdvantage® GM Atlantic salmon

non-GM Atlantic salmon of the same age

D The AquAdvantage salmon grow much faster than normal.

GM bacteria make a range of useful substances, including antibiotics and other medicines.

 6 Explain why some farmers want to grow GM crops.

 7 Milk contains lysozyme – an enzyme that prevents harmful bacteria growing in the intestines. Human breast milk contains 2500 times more lysozyme than cow's milk. Describe two ways of producing cows that make milk containing a lot of lysozyme.

 8 Explain why genetic engineering changes a species' genome but artificial selection does not.

Exam-style question

Scientists have created a goat that produces spider silk in its milk. Explain how the scientists would have done this. *(2 marks)*

Did you know?

Most cheese is made using enzymes produced by GM bacteria.

Checkpoint

How confidently can you answer the Progression questions?

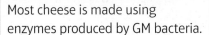 **Strengthen**

S1 People with haemophilia lack a blood protein called Factor VIII, so their blood does not clot properly. They can be treated with Factor VIII from donated blood, but this is expensive. Describe how another organism could be used to make Factor VIII more cheaply.

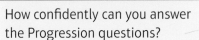 **Extend**

E1 Compare and contrast the use of selective breeding and genetic engineering in agriculture.

SB4f Tissue culture

Specification reference: B4.9B

Progression questions

- What is tissue culture?
- What are the advantages of using tissue culture in medical research?
- What are the advantages of using tissue culture in plant breeding?

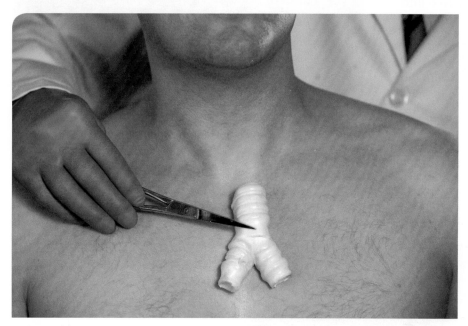

A The structure of a synthetic windpipe is created using finely spun, unreactive materials such as glass. The structure is then coated with stem cells from the patient to form new epithelial cells.

In 2011, Andemariam Beyene was the first person to receive a transplant of a tissue-engineered synthetic organ. Beyene's windpipe had been damaged by cancer, and he was given a new one made using **stem cells** from his bone marrow. This meant his immune system did not attack (**reject**) the new organ.

Tissue culture is the growing of cells or tissues in a liquid containing nutrients or on a solid medium (such as nutrient agar). This is a useful way to grow many identical cells. These may form a **callus** (a clump of undifferentiated cells). Sometimes the cells are then treated to make them **differentiate** (become specialised).

 1 Give one example of how tissue culture is used in medicine.

Tissue culture is used to produce new plants of very rare species which are at risk of **extinction**. It is also used to produce new individuals of plant species that may be difficult to grow from seed, such as orchids. The technique is also used to produce **clones** (identical copies) of GM plants.

 2 Explain why all the cells in a plant callus are genetically identical.

 3 Give one example of how tissue culture is used in plant breeding.

 4 Explain why sexual reproduction might not be as successful as tissue culture at making copies of a specific GM plant.

B Cells taken from an orchid plant are grown on solid agar to form a callus. Treatment with plant hormones will cause the callus cells to form the roots and shoots of new plants.

Did you know?

The Species Recovery Programme has used tissue culture to produce clones of many endangered species that grow in the East Himalayas. Some clones have been planted back in the wild.

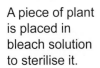

A piece of plant is placed in bleach solution to sterilise it.

Sometimes, a small piece of plant is cut off and placed on sterile nutrient medium to grow.

nutrient agar

The piece of plant is treated with hormones so it grows roots and shoots.

When the plants are large enough, they are planted into soil or compost.

Sometimes, only a few cells are cut off, and placed on sterile nutrient medium to grow into a callus.

The callus is treated with hormones so that plantlets develop with shoots and roots.

The plantlets are separated and grown on nutrient medium in sterile conditions.

C Everything must be sterile during tissue culture, to prevent the growth of microorganisms.

Tissue cultures have many uses in medicine. Culturing a thin layer of cells on a solid medium makes it easier to study how cells communicate with each other. Cell cultures are also needed to study **viruses**, which cannot replicate outside of cells. Cultures of cancer cells have been developed to study how cancers develop and spread. Using cell cultures, scientists can investigate how infected cells respond to new medicines without risking harm to animals or humans.

Cultures of human cells can be developed into tissues if correctly supported, as with the artificial trachea in photo A. A similar process is used to produce artificial bladders (part of the urinary system).

6 Describe how an artificial bladder for a specific patient might be produced using tissue culture.

7 Explain the advantages of using cell cultures as the first stage in developing a new medicine.

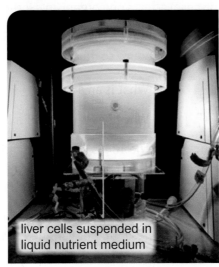

liver cells suspended in liquid nutrient medium

D It is not yet possible to make a fully functioning liver using tissue culture. However, a device that uses liver cells from animals can be used to support patients with liver damage while their own liver cells recover.

5 Explain why the nutrient medium (agar) and all the equipment are heated to a high temperature before they are used in tissue culture.

Checkpoint

How confidently can you answer the Progression questions?

Strengthen

S1 Design a summary table to show how tissue culture is used in plant breeding programmes, and the advantage of using this method compared with traditional breeding methods.

S2 Design a second table to summarise the use of tissue culture in medical research and its advantages.

Extend

E1 Vaccines are used to make people immune to a disease (so they do not fall ill). Vaccines against viruses can be made by treating real virus particles with a substance to make them harmless. Explain the advantages of using tissue culture to develop vaccines like this.

Exam-style question

Describe one advantage of using tissue culture in plant breeding and one advantage of using tissue culture in medicine. *(2 marks)*

Specification reference: **H** B4.11; B4.14

Progression questions

- What are the benefits and risks of selective breeding?
- What are the benefits and risks of genetic engineering?
- **H** How is genetic engineering carried out?

A Merino sheep have been bred to grow lots of fine wool. This one was found wearing a 40 kg fleece.

B Nearly half of all banana plants are one variety. A fungus is now threatening to destroy them.

Did you know?

Scottish scientist Helen Sang has produced GM chickens that lay eggs containing proteins that can be used to treat cancer.

Organisms are selectively bred or genetically engineered to grow faster, cope with environmental conditions, increase yields or make new products. However, there are drawbacks.

 1 Define the term 'GMO'.

Selective breeding risks

Genes exist in different forms, called **alleles**, which cause variation in characteristics. For example, the different alleles in wild wheat plants cause different seed sizes. In selective breeding, only certain alleles are selected. Others become rare or disappear. So, alleles that might be useful in the future are no longer available.

Farming huge numbers of the same breed or variety is also a problem. All the organisms are very similar and so if a change in conditions (e.g. a new disease) affects one organism, all the others are affected.

Animal welfare is a further concern. For example, some selectively bred chickens produce so much breast meat they can hardly stand up.

 2 **a** What is an allele?

 b Why does it matter if alleles are lost from a species?

 3 In the 1840s, many Irish families grew one variety of potato for food. Suggest why a disease (potato blight) caused mass starvation in Ireland at this time.

 4 Suggest why scientists now save the seeds of thousands of different plant varieties in seed banks.

Genetic engineering issues

GM crops have been produced to be resistant to some insects (so less insecticide is needed). Others are resistant to certain herbicides (weed killers) which then kill weeds but not the crop. These herbicides do not affect animals and are very effective against weeds, so less herbicide is used. However, the seeds for many GM plants are expensive. Some people think that GM crops will reproduce with wild plant varieties and pass on their resistance genes, and these genes may also have unknown consequences in wild plants. Others think that eating GM organisms may be bad for health (but there is not evidence to support this).

GM bacteria produce many useful substances, such as **insulin** (needed to treat **type 1 diabetes**). Insulin used to be extracted from dead pigs and cows, but insulin from GM bacteria is cheaper and suitable for vegans or people who do not eat pork or beef for religious reasons. However, it is slightly different to insulin from mammals and so not all diabetics can use it.

C Campaigners have destroyed GM crops during trials.

5 **a** State a benefit of Golden Rice (see *SB4e Breeds and varieties*).

b Suggest why some people are against growing Golden Rice.

6 Suggest a problem of wild plants becoming resistant to a herbicide.

H Genetic engineering of bacteria

A bacterium has one large loop of DNA (containing most of its genes) and some small circles of DNA, called **plasmids**. To genetically engineer bacteria, additional genes are added to a plasmid. The plasmid is made of DNA combined in a new way and so it is an example of **recombinant DNA**.

Scientists use **restriction enzymes** to cut a useful gene out of an organism's DNA. This cutting leaves strands of DNA with jagged ends, called **sticky ends**. If two sticky ends match, they can be joined together using an enzyme called **ligase**. Diagram D shows the process.

Any DNA molecule used to carry new DNA into another cell is called a **vector**.

1 Restriction enzymes make staggered cuts in DNA molecules, producing sections with a few unpaired **bases** at each end – 'sticky ends'. A section of DNA containing the gene for making insulin is cut from a human chromosome in this way.

2 Restriction enzymes are also used to cut plasmids open. By using the same restriction enzyme as was used on the human chromosome DNA, the cut plasmids have the same sticky ends.

3 Sections of DNA containing the insulin gene are mixed with the cut plasmids. The complementary bases on the sticky ends pair up. An enzyme called ligase is used to join the ends together.

4 The plasmids are then inserted back into bacteria, which are then grown in huge tanks. The insulin they now make can easily be extracted.

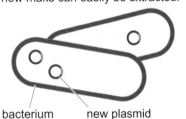

bacterium   new plasmid

D genetic engineering of a bacterium

7 **a** Draw a flow chart to describe how to genetically engineer a bacterium.

b What vector is used in this process?

Checkpoint

How confidently can you answer the Progression questions?

Strengthen

S1 Discuss how a farmer might decide whether or not to plant a large area with one variety of wheat that is suited to that area.

Extend

E1 **H** Explain the importance of using just one type of restriction enzyme in genetic engineering.

Exam-style question

Describe the advantages of making insulin using genetically modified bacteria rather than extracting it from animals. *(2 marks)*

Progression questions

- How can crop plants be modified to make them resistant to insects?
- What are the advantages of producing GM organisms?
- What are the disadvantages of producing GM organisms?

A Large areas of one kind of crop are called monocultures. The crop can provide so much food for insects that their numbers rapidly increase and they become a pest.

In 2014, over 78 million tonnes of maize (sweetcorn) were grown in Europe. This crop provides food for us, farm animals and a lot of insects (such as aphids).

When insects eat a crop, they can damage it and reduce its **yield** (the amount of the crop we can use). These insects are **pests**. A pest of maize is the European corn borer (photo B), whose caterpillars can reduce the yield by over 10 per cent.

Insect pests can be controlled by spraying the crop with chemical **insecticides**. Different insecticides kill different insects, and many only affect an insect when they touch it.

 1 Explain why some insects are a problem in agriculture.

 2 Suggest why the European corn borer is particularly difficult to kill with insecticide.

In the early 1900s, a soil bacterium (*Bacillus thuringiensis*) was discovered that makes a natural insecticide protein called **Bt toxin**. Crystals of the toxin can be sprayed onto crops as insecticide. In 1985, the genes that control the production of Bt toxin in the bacterium were introduced into plants, so that all cells in the plants produced the toxin.

 3 Suggest how plants were genetically modified to produce Bt toxin.

 4 Explain why the Bt toxin in a GM maize crop can target the European corn borer, when spraying plants with Bt toxin crystals cannot.

One advantage of GM Bt toxin is that it only affects insects that chew the plant tissues, as the toxin is released when the cells are broken, while insecticide sprays may kill a wide range of insect species. Insect predators, such as ladybirds and spiders, are unharmed by GM maize because they do not eat the plants. However, insect *pests* that suck sap from the plant, such as aphids, do not chew plant tissue either and so do not eat the toxin. Farmers may still have to spray their crops with insecticides that control these other pests.

B Once the caterpillar (larva) of a European corn borer has burrowed into a maize stalk, it will live there for several months, eating the inside of the plant.

Another problem with growing crop plants that make their own insecticide is that the insects can develop **resistance** to the toxin, which means it no longer harms them. Fortunately, there are many different **strains** (varieties) of *B. thuringiensis* bacteria, which produce slightly different forms of the toxin. New versions of the GM crop plants can be developed to replace the varieties that the pests are resistant to.

Many different GM crops have been produced that contain new genes that make them resistant to a range of pests as well as to some herbicides (weedkillers). GM crop seeds are more expensive than non-GM varieties, but farmers usually make more profit by growing the GM varieties – as long as people are willing to buy food from GM crops.

Some people are concerned that eating GM foods could harm their health, although there is currently no evidence to support this claim. Others worry that the new genes might transfer to other crops or to wild plants by pollination, but research suggests this happens very rarely.

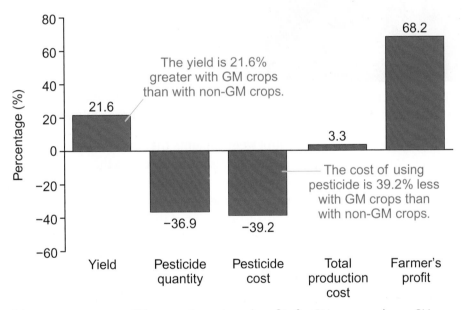

C Average percentage differences in costs and profits for GM crops and non-GM crops. A positive value means GM crops have a greater value than non-GM.

 6 **a** Use graph C to identify advantages and disadvantages of growing GM crops instead of non-GM varieties.

b Give reasons for the differences in values between GM and non-GM crops for each factor shown in graph C.

c Suggest how development of insect resistance to Bt toxin would alter the values in chart C.

Exam-style question

Describe how a crop plant could be genetically modified with the Bt gene to make it resistant to an insect pest. *(3 marks)*

 5 **a** Explain why using an insect-resistant variety of GM maize might be better for the environment than spraying non-GM maize with insecticide.

 b Aphids breed rapidly in the summer. Suggest why aphid pests may be found in greater numbers on a crop that produces Bt toxin compared with one that does not.

Did you know?

The first plants to be genetically modified with Bt toxin were tobacco plants. These plants are easy to grow and easy to modify genetically.

Checkpoint

How confidently can you answer the Progression questions?

Strengthen

S1 Explain how growing GM crops that are insect-resistant could:

a benefit the environment

b harm the environment.

Extend

E1 A new GM crop plant has been developed to be resistant to aphid pests. Explain how the GM variety should be tested before it can be released for farmers to buy.

SB4i Fertilisers and biological control

Specification reference: B4.13B

Progression questions

- What is biological control?
- What are the advantages and disadvantages of using biological control?
- What are the advantages and disadvantages of using fertilisers on crops?

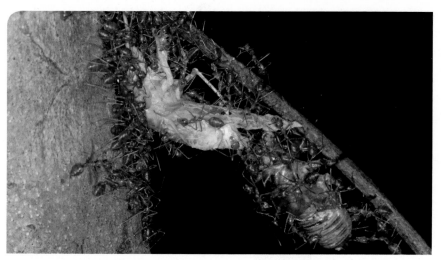

A Weaver ants work as a team to attack and kill large prey, such as this newly hatched cicada. Cicadas are cricket-like insects that suck plant sap.

Using insecticide is not the only way to control insect pests. For over 1000 years, Chinese farmers have placed weaver ant nests in their citrus trees. The predatory ants remove some insect pests, and so help to increase crop yield. Using organisms to control pests is known as **biological control**.

 1 a Explain why the use of weaver ants is an example of biological control.

 b Explain how using weaver ants can help citrus farmers grow more oranges.

B *Chrysolina* beetles are herbivores that prefer eating St John's Wort to many other plant species.

Biological control can also control **weeds** (plants that cause problems). St John's Wort became a weed when it was introduced into the USA. It grows well in grassland but can cause illness in farm animals that eat it. Instead of using herbicides over a large area, *Chrysolina* beetles were introduced. Within 10 years, there was less than 1 per cent of the original amount of the weed left in California, where it had been particularly damaging.

 2 a Explain why *Chrysolina* beetles were useful for biological control of St John's Wort.

 b Before the beetles were released into the wild, lab tests were done to check what the beetles ate. Suggest why this was done.

Did you know?

Some attempts at biological control turn out to be disasters. The Asian harlequin ladybird was introduced to orchards in the USA to eat aphid pests. The ladybirds were so good at this that the native predators of aphids soon had no food. Even worse, the ladybirds also ate other native insects.

Biological control, GM organisms and selective breeding can all help to increase the amount of food we produce. However, as the human population continues to grow, we need to use all the methods we have to increase food production. This includes using **fertilisers** to increase the growth and yield of crop plants.

Mineral salts are naturally occurring compounds found in rocks and soils. Plants need ions from these compounds to produce new substances. Fertilisers contain mineral ions, such as nitrogen, potassium and phosphorus, that plants absorb from the soil to make healthy new cells. More fertiliser needs to be added with each new crop so that it will grow well.

 3 a Explain which test plot in the field in photo C did not get fertiliser.

 b Explain why adding fertiliser affects the yield of crop plants.

 4 Explain why more fertiliser must be added to a field with each new crop.

If not all the fertiliser is absorbed by a crop, some may get into nearby streams, rivers and lakes. This can cause **pollution** and lead to the death of organisms in the water. It can also cause health problems for humans and animals if they drink the water.

C After the test plots in this field were planted, some were not given fertiliser containing nitrogen.

D Nitrogen compounds called nitrates are particularly important mineral salts for plant growth. However, too much nitrate in food or water can cause blue-coloured membranes in the mouth and even the death of farm animals, such as these cattle in Dabhla village, India. It is also harmful for humans, particularly babies.

 5 Explain why the amount of fertiliser added to a field must be carefully measured.

 6 Fertilisers are highly soluble. Explain why farmers should check the weather forecast before spraying a crop with fertiliser.

Exam-style question

Describe one advantage and one disadvantage of using fertilisers on crops.
(2 marks)

Checkpoint

How confidently can you answer the Progression questions?

Strengthen

S1 A farmer has a crop of cabbages that are being eaten by caterpillars. Suggest how biological control could help the farmer, and what problems this might cause.

Extend

E1 Before potential biological control organisms can be released into the environment, they must be approved. Suggest the criteria that would be used to judge whether an organism is suitable for biological control.

Genetic engineering

The European corn borer damages maize crops in the south of the UK. Discuss whether farmers should be allowed to grow genetically modified corn that is resistant to this insect in the UK.

(6 marks)

Student answer

There are advantages and disadvantages to GM corn [1]. The GM corn is resistant to European corn borers [2]. This GM corn is good because it kills a pest and so less pesticide is needed, but there are other pests and so some pesticide will still be needed [3]. Some people think GM will be bad for them [4].

[1] For a discussion question, first think about the problem that is being addressed in the question. This is a good start since there are pros and cons that need to be thought about when reaching a decision.

[2] Avoid repeating information given in the question.

[3] Excellent point, stating an advantage and explaining why it is an advantage and then going on to balance that up with a disadvantage.

[4] Another disadvantage, but why people think GM will be bad for them is not explained (e.g. may cause allergies).

Verdict

This is an acceptable answer. It gives both an advantage and a disadvantage of the GM corn.

It would be better if there were a few more advantages and disadvantages, and if there was more scientific detail. For example, the answer could contain more detail on why people do not want to eat GM crops. The answer talks about the fact that less pesticide is needed if the GM crops are grown, but it could also then go on to explain that pesticides can be harmful. This would show that the student can link scientific ideas.

Exam tip

A question that says 'Discuss …' really means 'What do you think about … ?'. Write down a couple of points 'for' and a couple of points 'against'. Then build an argument around these. Don't be scared to say what you think, but you need to back up your ideas with scientific information.

Paper 1

SB5 Health, Disease and the Development of Medicines

Hidekichi Miyazaki set a record of 42.22 seconds for the 100 m sprint at the age of 105. Miyazaki put his success down to daily exercise and eating healthily.

In the UK, between 2002 and 2012 there was a 73 per cent increase in the number of people living beyond 100 years. There are similar increases in other parts of the world. Although governments are concerned that the cost of looking after older people will also increase, people are living healthily and are active to a much older age than before. Maybe one day we will see sporting competitions just for those over 100.

The learning journey

Previously you will have learnt at KS3:

- that imbalances in the diet can lead to obesity and deficiency diseases
- that recreational drugs (such as alcohol) can affect behaviour, health and life processes.

You will also have learnt in *SB1 Key Concepts in Biology*:

- about the structure of bacteria
- about the use of microscopes to study cells.

In this unit you will learn:

- about how we define health
- about some pathogens, the diseases they cause, and how their spread can be reduced or prevented
- about the lifecycle of viruses
- how plants defend themselves from pests and pathogens
- how the body is protected against infection
- about the immune system
- how antibiotics work
- about aseptic techniques for culturing microorganisms
- how new medicines are developed
- **H** how plant diseases can be identified.

Specification reference: B5.1; B5.2; B5.3

Progression questions

- What is health?
- How do communicable and non-communicable diseases differ?
- Why can having one disease increase the chance of getting another disease?

A Physical fitness improves physical well-being. Exercising as part of a group can also improve your social and mental well-being.

 1 Use your own words to define the term 'good health'.

The World Health Organization (WHO) is responsible for coordinating ways to improve **health** across the world. According to the organisation, good health means more than simply feeling well; it is a state of 'complete physical, social and mental well-being'.

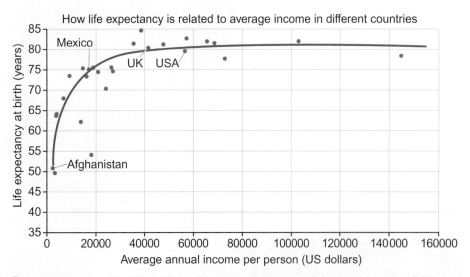

B Graph showing a link (correlation) between income and the average age of death (life expectancy) for people living in different countries.

- Physical well-being includes being free from **disease**, eating and sleeping well, getting regular activity, and limiting the intake of harmful substances such as alcohol and drugs.

- Social well-being includes how well you get on with other people, and also how your surroundings affect you.

- Mental well-being includes how you feel about yourself.

These three categories of well-being are not separate – improving one category can also improve the others.

Graph B shows a **correlation** between health and income: the smaller someone's income, the less likely they are to live for a long time. However, we cannot be entirely sure of the **cause** of this correlation. It may be because poorer people cannot afford as healthy a diet, or access to the same medical care, as those with more money. In regions where there are disasters such as floods or wars, people are also more likely to have poor health because it is likely there will be more disease and fewer doctors and hospitals.

 3 Explain what is meant by a correlation between two factors.

 4 a Describe what the shape of the curve in graph B shows.

 b Suggest a reason for the correlation shown in graph B. Explain your answer.

A disease is a problem with a structure or process in the body that is not the result of injury. A disease might be due to microorganisms getting into the body and changing how it works. For example, the flu virus can cause a high temperature, sneezing and aches and pains. Microorganisms that cause diseases are called **pathogens**. Diseases caused by pathogens are **communicable diseases**, as they can be passed from an infected person to other people.

Some diseases are **non-communicable**, because they are not passed from person to person. They are caused by a problem in the body, such as a fault in the genes (as in cystic fibrosis) or as a result of the way we live – our **lifestyle**.

Diseases may be correlated, so that having one disease means a person is more likely to have another disease. Possible causes of these correlations include:

- one disease damages the **immune system**, making it easier for other pathogens to cause disease (the immune system protects the body from communicable diseases; one pathogen that attacks the immune system is the HIV virus)
- a disease damages the body's natural barriers and defences, allowing pathogens to get into the body more easily
- a disease stops an organ system from working effectively, making other diseases more likely to occur.

 5 a Periodontal disease is correlated with heart disease. Describe this correlation.

 b Suggest a cause of this correlation.

2 Suggest how exercising regularly as part of a group can improve:

 a your physical well-being

 b your social well-being

 c your mental well-being.

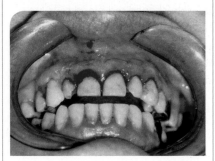
Checkpoint

How confidently can you answer the Progression questions?

Strengthen

S1 Suggest why somebody infected with the HIV virus is more likely than people without the virus to get other communicable diseases.

Extend

E1 Suggest why where you live might affect how long you live for.

Exam-style question

Describe the difference between communicable and non-communicable diseases. Use an example of each in your description. *(2 marks)*

SB5b Non-communicable diseases

Specification reference: B5.23; B5.24

Progression questions

- What do non-communicable diseases have in common?
- How can diet affect malnutrition?
- Why does alcohol cause problems for people and for society?

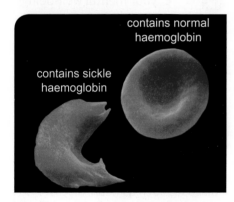

A Sickle-shaped red blood cells are caused by a mutation in a gene that codes for haemoglobin. A person with two sickle mutant alleles suffers from sickle cell disease.

 2 Suggest how scurvy should be treated. Explain your answer.

3 Suggest why kwashiorkor is usually only seen in people living in very poor parts of the world.

 4 Explain why deficiency diseases are examples of malnutrition.

Did you know?

Rickets used to be thought of as a disease of the past in the UK, but studies show it is increasing again, particularly in the under-fives. A poor diet may be only part of the problem. Rickets can also be caused by a lack of sunlight because vitamin D is produced naturally in the skin when in sunshine.

There are several different types of non-communicable disease. One type is **genetic disorder** (genetic disease) caused by faulty alleles of genes. Genetic disorders can be passed to offspring but not to any other person.

 1 Explain why sickle cell disease is a non-communicable disease.

Other non-communicable diseases occur as a result of poor diet or **malnutrition**. Malnutrition occurs when you get too little or too much of particular nutrients from your food. The lack of a certain nutrient can cause a specific **deficiency disease**.

Nutrient	Disease caused by deficiency of nutrient	Symptoms of disease	Good sources in diet
protein	kwashiorkor	enlarged belly, small muscles, failure to grow properly	meat, fish, dairy, eggs, pulses (e.g. lentils)
vitamin C	scurvy	swelling and bleeding gums, muscle and joint pain, tiredness	citrus fruits (e.g. oranges) and some vegetables (e.g. broccoli)
vitamin D and/or calcium	rickets or osteomalacia	soft bones, curved leg bones	vitamin D: oily fish calcium: dairy products
iron	anaemia	red blood cells that are smaller than normal and in reduced number, tiredness	red meat, dark green leafy vegetables, egg yolk

B some diseases caused by lack of particular nutrients

Alcohol and disease

Some diseases are caused by how we choose to live – our lifestyle. This includes whether we take enough exercise and whether we take drugs. Ethanol, found in alcoholic drinks, is a **drug** because it changes the way the body works. Ethanol is broken down by the liver, and a large amount of ethanol taken over a long period can lead to liver disease, including **cirrhosis**.

A cirrhotic liver does not function well and can result in death. Liver disease is the fifth largest cause of death in the UK. Deaths from alcohol-related liver disease have increased by 450 per cent in the last 30 years in the UK. The cost of treating people with liver disease is more than £500 million each year and is still rising.

5 Explain why liver disease is a non-communicable disease.

6 Explain why there is a Drink Awareness campaign in the UK aimed at limiting the amount of alcohol people drink.

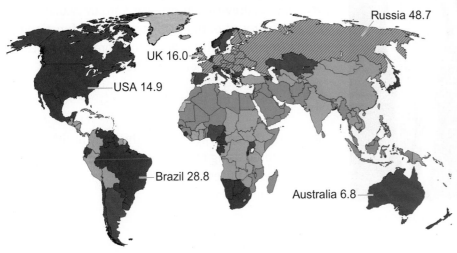

UK 16.0

USA 14.9

Russia 48.7

Brazil 28.8

Australia 6.8

Alcohol (dm³) per person per year

- <2.50
- 2.50–4.99
- 5.00–7.49
- 7.50–9.99
- 10.00–12.49
- ≥12.50
- Data not available

D Global map of alcohol consumption in different countries. Death rates (number per 100 000 people) from liver disease are labelled for five countries.

7 Look at map D.

  a Put the named countries in order (starting with the highest value) of:

   i deaths from liver disease

   ii consumption of alcohol per year.

  b What do your lists from part **a** suggest about the correlation between deaths from liver disease and alcohol consumption? Explain your answer.

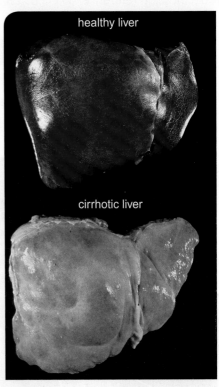

healthy liver

cirrhotic liver

C A healthy liver is dark red, smooth and soft. A liver that has cirrhosis may be paler and larger, rough and much harder.

Checkpoint

How confidently can you answer the Progression questions?

Strengthen

S1 Give one reason why too much alcohol over a long time is a problem for each of the following.

- the person who drinks it
- their family
- the society they live in

Extend

E1 The UK Department of Health recommends that all children from six months to five years should take vitamin D drops every day to supplement their diet. Discuss the advantages and disadvantages of this advice.

Exam-style question

Explain how a person's diet can cause anaemia. *(2 marks)*

SB5c Cardiovascular disease

Specification reference: B5.24; B5.25

Progression questions

- What is cardiovascular disease?
- What effect do smoking and obesity have on the risk of developing cardiovascular disease?
- Why are there a range of treatments for cardiovascular disease?

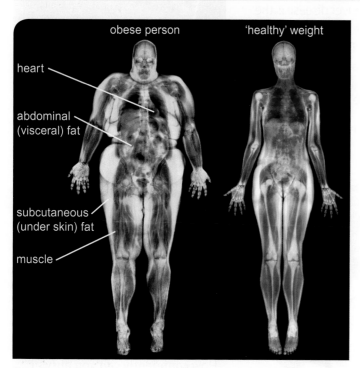

A the tissues and organs of an obese person and a person of healthy mass ('weight')

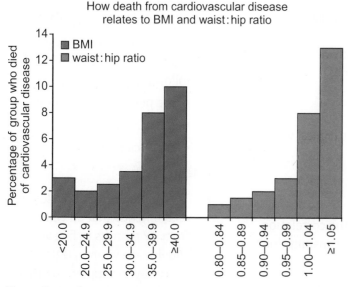

B correlation between BMI and waist:hip ratio with death caused by cardiovascular disease for over 4000 Australian men

Malnutrition caused by a diet that is high in sugars and fats can lead to **obesity**, where large amounts of fat are formed under the skin and around organs such as the heart and kidneys.

We need some fat to cushion organs when we move, to store some vitamins, and to provide a store of energy. Too much fat can increase the risk of many diseases, including **cardiovascular disease**. Cardiovascular disease is a result of the circulatory system functioning poorly. One sign of this is high blood pressure, which can lead to heart pain or even a **heart attack**.

 1 Define the term 'obesity'.

Measuring the amount of fat on the body is difficult, so it is estimated using other measures. **Body mass index (BMI)** for adults uses height (in metres) and mass (in kilograms) in the formula:

$$\text{BMI} = \frac{\text{mass}}{\text{height}^2}$$

When BMI is used to predict the amount of fat, it assumes that the mass of other body tissues is in proportion to height. An adult who has a BMI of 30 or more is usually considered obese.

 2 Explain why a BMI of 30 or more is used as a predictor of cardiovascular disease.

 3 Explain why BMI is not a good predictor of body fat for a weightlifter with large muscles.

The fat that seems to be most closely linked with cardiovascular disease is abdominal fat. Dividing waist measurement by hip measurement to get **waist-to-hip (waist:hip) ratio** gives a better method of measuring abdominal fat than BMI.

4 Use the evidence in chart B to explain why waist:hip ratio is better than BMI when looking at death caused by cardiovascular disease.

Smoking and disease

Tobacco smoke contains many harmful substances that can damage the lungs when they are breathed in. Some of these substances are absorbed from the lungs into the blood and are transported around the body. These substances can damage blood vessels (diagram C), increase blood pressure, make blood vessels narrower and increase the risk of blood clots forming in blood vessels. All of these can lead to cardiovascular disease.

Substances from tobacco smoke damage the artery lining.

Fat builds up in the artery wall at the site of damage, making the artery narrower.

A blood clot may block the artery here, or break off and block an artery in another part of the body – causing a heart attack or **stroke**.

C Damage to blood vessels by substances from tobacco smoke can cause the build-up of fat in an artery.

Treating cardiovascular disease

High blood pressure increases the risk of cardiovascular disease. A doctor may advise a patient with high blood pressure to exercise more and give up smoking. If blood pressure is very high, then the patient may be given medicines to reduce it.

A narrowed blood vessel can be widened by inserting a small mesh tube (**stent**) at the narrowest part to hold it open. Blocked arteries in the heart can be bypassed by inserting other blood vessels so that the heart tissue is supplied with oxygen and nutrients again. Patients who have these operations may have to take medicines for the rest of their lives to help prevent a heart attack or stroke.

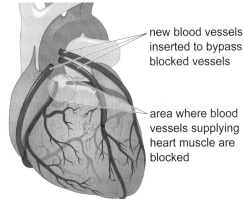

new blood vessels inserted to bypass blocked vessels

area where blood vessels supplying heart muscle are blocked

D Blocked blood vessels can be bypassed by inserting new blood vessels.

 5 Sketch a flowchart to show how smoking can lead to cardiovascular disease.

 6 Heart muscle has its own blood supply through coronary arteries. Explain why narrowed or blocked coronary arteries must be bypassed to avoid cardiovascular disease.

Checkpoint

How confidently can you answer the Progression questions?

Strengthen

S1 Explain why a doctor may advise a patient with a high BMI to give up smoking and to exercise more.

Extend

E1 Explain why 'prevention is better than cure' is a good approach to the problem of cardiovascular disease.

Exam-style question

State and explain two methods of treating cardiovascular disease. *(2 marks)*

Progression questions

- How can pathogens spread?
- What is a vector of disease?
- How can the spread of pathogens be reduced or prevented?

A The 'catch it, bin it, kill it' slogan encourages people to use tissues and wash their hands when infected with a disease such as flu. The idea is to stop the spread of many pathogens that are spread through the air.

 2 Explain why it was difficult to prevent the spread of chalara ash dieback into the UK.

Did you know?

At the time that Pacini identified the cholera bacterium, John Snow was mapping the outbreaks of cholera in parts of London. The map he produced showed how cholera infections and water supply were linked.

Most pathogens cannot grow outside their host, and so must spread from one host to another so they can increase in number. If we know how pathogens are spread, then we can find ways of stopping that spread.

Infections such as colds, flu and tuberculosis (TB) cause a person to sneeze or cough. This sends droplets containing pathogens into the air. Once in the air, flu viruses can survive for about a day. However, TB bacteria can survive for months in air, and mix with dust that can blow around and infect another person. Fungi, such as that causing chalara ash dieback, can also spread in the air, as tiny tough spores (cells that can grow into new organisms). Strong winds can carry chalara spores over long distances, such as to the southern UK from nearby Europe.

 1 a Look at photo A. Explain how the 'catch it, bin it, kill it' idea could help to reduce the spread of TB bacteria.

 b Suggest one other method that could be used to reduce the number of TB bacteria in the environment.

Some pathogens spread in water, such as the bacteria that cause cholera, typhoid and dysentery (which all cause severe diarrhoea). These diseases are normally rare in developed countries, because the water that we use for drinking, cooking and washing is treated to kill pathogens. Keeping things clean to remove or kill pathogens is known as good **hygiene**. Outbreaks of these diseases occur when hygiene is difficult, such as in very poor areas, after major environmental disasters, or in refugee camps.

B After a major earthquake in Haiti in 2010, pipes carrying clean water were broken, and people had to get their drinking water from polluted wells and rivers.

Pathogens of the digestive system can spread in food as well as water. They enter the body through the mouth, which is described as the **oral route**. *Helicobacter* bacteria are thought to be spread when people touch other people's food after touching their mouths, or after going to the toilet (oral-faecal transmission). The bacteria may also spread on the feet of flies that have fed on infected faeces and then landed on food.

 4 Suggest two ways in which the spread of *Helicobacter* could be reduced, and explain your answers.

A few pathogens, such as the Ebola virus, require extreme hygiene practices to control them. The virus very easily enters people's bodies through broken skin or the eyes, nose or mouth. The 2014–15 Ebola outbreak, which mainly affected West Africa, became an **epidemic** because many people became infected when burying those who had died from Ebola.

 5 Explain why normal hygiene practices, such as cleaning hands thoroughly, do not prevent the spread of Ebola.

Some pathogens cannot survive in the environment and so must spread in other ways. For example, the malaria protist is carried in blood by mosquitoes that sucked blood from an infected person. The mosquito injects the protist directly into the blood of the next person it feeds on. Organisms that carry pathogens from one person to the next are called **vectors** of disease. Controlling the spread of the pathogen may involve controlling the spread of the vector.

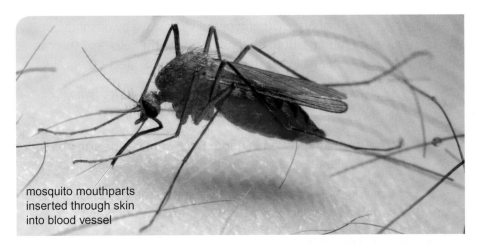

mosquito mouthparts inserted through skin into blood vessel

D Female *Anopheles* mosquitoes feed on human blood by piercing the skin with their mouthparts. The blood may also carry malaria protists.

 6 Explain why killing mosquitoes could help control the spread of malaria.

Exam-style question

Explain two ways that the spread of cholera in a refugee camp could be prevented. *(2 marks)*

 3 a Suggest why cholera spread after the 2010 Haiti earthquake.

b Suggest hygiene practices that could have prevented this spread of cholera.

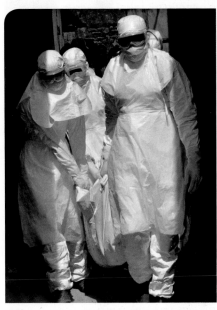

C The best way to avoid infection with the Ebola virus is by wearing full body protection, because the virus is present in all body fluids of infected people, even after death.

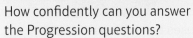

Checkpoint

How confidently can you answer the Progression questions?

Strengthen

S1 Explain why it is important to wash your hands thoroughly after going to the toilet.

Extend

E1 Explain how isolating infected people and wearing full-body protective clothing helped to bring the 2014–15 Ebola epidemic under control.

SB5f Virus life cycles

Specification reference: B5.7B; B5.19B

Progression questions

- What is a virus?
- What happens in the lytic and lysogenic pathways of a virus' life cycle?
- How can we compare the effects of viruses?

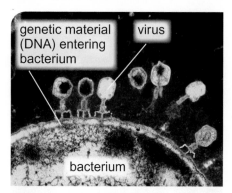

A Viruses that attack bacteria attach to the surface of a bacterium before injecting their genetic material into the cell (magnification ×55 000).

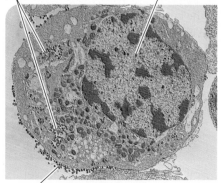

B A lymphocyte (white blood cell) infected with HIV. Viruses outside the cell have pushed out through the cell membrane (magnification ×47 500).

There are many different types of virus, but they all have certain features in common. All viruses contain one or more strands of genetic material surrounded by a protein coat, or **capsid**, although many have additional layers surrounding the capsid.

All viruses are unable to replicate (make copies) on their own. They have to enter a living cell and take over that cell's processes for making new genetic material and proteins. Different types of virus take over different types of cell – some invade plant cells, others invade bacterial cells. Viruses that take over human cells include HIV (human immunodeficiency virus) and Ebola virus.

 1 Describe the basic structure of a virus.

 2 Explain why viruses cannot replicate outside a living cell.

The cell copies the viral genetic material and makes new viral genetic material and proteins. These components assemble into new viruses, which then escape from the cell. Some types of virus cause the complete breakdown of the cell, or **lysis**. Other types of virus leave by pushing out through the cell membrane. Both methods damage the cells, and this causes disease. The viruses that have left the cell can go on to infect other cells.

 3 a Use the magnification in photo A to estimate the size of one virus.

 b Use the magnification in photo B to estimate the size of a white blood cell and one of the viruses.

 c Use your answers to **a** and **b** to compare the size of the viruses with that of a bacterial cell and a human white blood cell.

 4 Explain why viruses can cause disease.

Viruses that cause lysis of a cell go through a **lytic pathway** during their life cycles. Sometimes when they infect a cell, these viruses behave differently. Their genetic material inserts into the cell's genetic material. Every time the cell divides, the virus' genetic material is replicated with the cell's genetic material. This may happen many times. This is known as the **lysogenic pathway**. At some point, the virus' genetic material triggers the copying of itself and the making of viral proteins, and the virus returns to its lytic pathway.

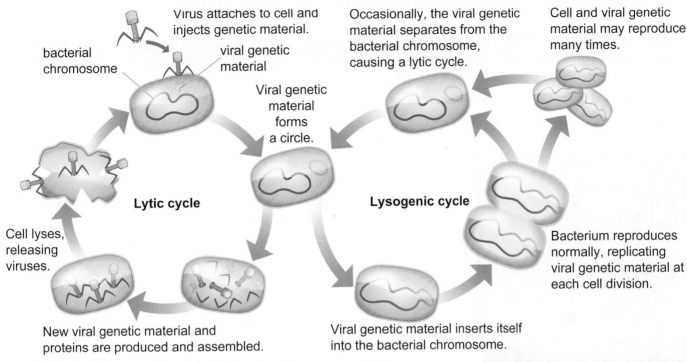

bacterial chromosome

Virus attaches to cell and injects genetic material.

viral genetic material

Occasionally, the viral genetic material separates from the bacterial chromosome, causing a lytic cycle.

Cell and viral genetic material may reproduce many times.

Viral genetic material forms a circle.

Lytic cycle

Lysogenic cycle

Cell lyses, releasing viruses.

Bacterium reproduces normally, replicating viral genetic material at each cell division.

New viral genetic material and proteins are produced and assembled.

Viral genetic material inserts itself into the bacterial chromosome.

C the lytic and lysogenic pathways in the life cycles of some types of virus

 5 Describe how viral genetic material is replicated in the lytic and lysogenic pathways of a virus' life cycle.

The effect of viruses on bacteria can be studied using **bacterial lawn plates**. These plates are made with **nutrient agar**, on top of which a thin layer of bacteria grows. A solution containing viruses is added to the plate. After a day or two, clear circles can be seen where bacteria have been killed by the viruses. The **cross-sectional area** of a clear circle is calculated using the equation:

cross-sectional area = πr^2 (where r is the radius of the circle)

The larger the clear area, the more effective the viruses have been at replicating and killing the bacteria.

 6 a A bacterial plate was infected with viruses. After a few days the radius of clear space A was 1.2 mm and the radius of clear space B was 1.4 mm. Calculate the cross-sectional areas of the clear spaces.

 b Explain which area contained the viruses that replicated faster and killed more bacteria.

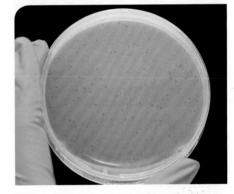

D The clear areas on this bacterial lawn plate show where viruses have killed the bacteria.

Checkpoint

How confidently can you answer the Progression questions?

Strengthen

S1 Describe the main stages in the life cycles of all viruses.

Extend

E1 Explain why testing the activity of a virus on a bacterial lawn plate would not be suitable for a virus that has a long lysogenic part in its life cycle.

Exam-style question

Compare the lytic and lysogenic pathways in a virus' life cycle.   *(2 marks)*

SB5g Plant defences

Specification reference: B5.9B; B5.10B; B5.17B

Progression questions

- How do plants protect themselves using physical barriers and chemical substances?
- How do we use some of the substances that plants make to protect themselves?
- Why is aseptic technique important when testing the activity of plant substances on bacteria?

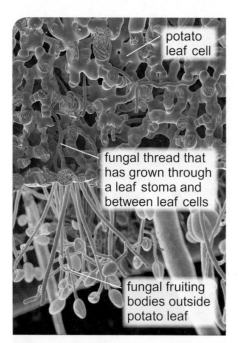

potato leaf cell

fungal thread that has grown through a leaf stoma and between leaf cells

fungal fruiting bodies outside potato leaf

A Fungal pathogens of plants release enzymes from the tip of each thread to help them get to nutrients inside the cells (magnification ×200).

Plants protect themselves from attack by pathogens such as bacteria, fungi and viruses. The outer surfaces of leaves and stems are covered by a waxy layer called the **cuticle**. This layer acts as a **physical barrier**, making it difficult for pathogens to get to the cells beneath. Woody plants, such as bushes and trees, also protect their stems with a thick layer of bark.

If pathogens get through the barriers, they must then penetrate the tough cell walls to get inside the cells. Some pathogens do this by releasing enzymes that soften cell walls. Others infect parts of plants that have weaker cell walls (such as young shoots and parts of plants that are not growing well).

 1 Give a reason why the cuticle is a physical barrier to infection.

 2 Explain how some fungi overcome plant defences to infect the plant.

 3 Explain why plants that have physical damage, such as a wound or a cut in the surface, are more easily infected by pathogens than undamaged plants.

Physical barriers are not good protection against herbivores, including **pests** such as caterpillars and aphids. Instead many plants use **chemical defences** such as poisons or insect repellents. Some plants produce chemical substances to deter herbivores and pathogens only when they are attacked.

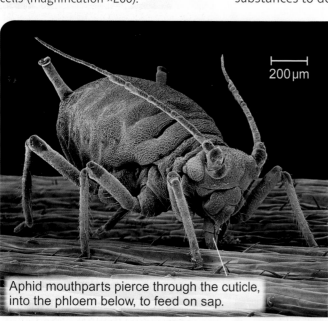

Aphid mouthparts pierce through the cuticle, into the phloem below, to feed on sap.

For example, one type of wild potato releases a substance into the air when attacked by aphids. This substance is very like the alarm substance that aphids release when attacked by a predator, and causes other aphids to fly away. Other plants, such as foxgloves, produce poisons all the time. Producing these substances takes energy, so it must be worth the cost to the plant to make them.

 4 Give two reasons why a gardener may be worried by seeing aphids on plants.

 5 **a** Explain how the wild potato protects itself against aphids.

 b Suggest one advantage to the potato plant of only producing a substance when it is attacked.

B Aphid mouthparts often transfer plant pathogens, such as viruses, between the plants that the aphids feed on.

C Foxgloves produce a substance that affects the heart rate of herbivores such as sheep.

 6 Suggest why foxgloves make poisons all the time.

Many medicines that we now use were developed from substances that plants use to protect themselves. For example, aspirin is commonly used to control **symptoms** of pain or fever. Aspirin was originally produced from salicylic acid, which is made by several plants, including meadowsweet and willow trees. Another example is the medicine artemisinin, which kills the *Plasmodium* protists that cause malaria. It was originally extracted from the wormwood plant, which has been used for centuries in Chinese herbal medicines to treat malaria.

 7 Describe two ways in which substances from plants are used medically.

Today, most new medicines are produced using chemical substances in a laboratory. During development, new medicines are tested on cultures of bacteria or human cells. These tests must not become contaminated by microorganisms from the air and on equipment. A series of **aseptic techniques** are used, including the use of an **autoclave** to **sterilise** equipment and growth medium (e.g. the nutrient agar that is used to grow bacteria in Petri dishes).

 8 a Give two examples of aseptic techniques.

 b Explain why aseptic techniques are important when testing substances on cells.

Exam-style question

Explain why cultures of bacteria in Petri dishes and vials must be kept covered.

(2 marks)

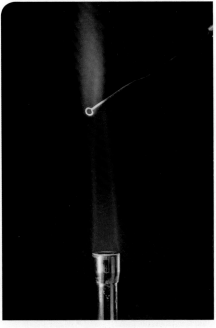

D An inoculating loop used for transferring microorganisms between cultures can be sterilised by 'flaming' it (making it glow) in a hot Bunsen flame.

Checkpoint

How confidently can you answer the Progression questions?

Strengthen

S1 Young ash trees have thinner leaf cuticles and cell walls than older ones. Explain how this helps to explain why they are more likely to be infected by chalara ash dieback than older trees.

Extend

E1 Explain why healthy plants that withstand attack by pathogens often become infected when they lack water or nutrients.

SB5h Plant diseases

Specification reference: **H** B5.11B

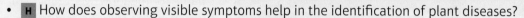

Progression questions

- **H** How does observing visible symptoms help in the identification of plant diseases?
- **H** How does distribution analysis help in the identification of plant diseases?
- **H** How does diagnostic testing help in the identification of plant diseases?

H

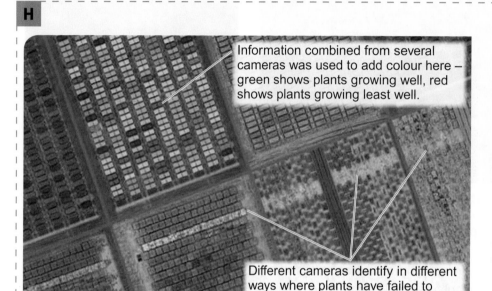

Information combined from several cameras was used to add colour here – green shows plants growing well, red shows plants growing least well.

Different cameras identify in different ways where plants have failed to grow due to drought or waterlogging.

A These field plots were photographed at different times by a drone using different kinds of camera. The different colours of plots within a field are used to identify where crops are growing poorly due to disease or drought.

Cameras on drones are a powerful tool for farmers, helping them to find out if crop plants are under stress. Plants show signs of stress whenever conditions are not good for growth, such as when there is too much or too little water, when the soil lacks nutrients, or when the plants are attacked by **pests** or diseases. Sometimes the stress may be caused by a combination of factors. Identifying the cause of stress is essential so that the farmer can treat the crop correctly and prevent loss of **yield** (the amount of useful product).

 1 Suggest why crop plants that have diseases may give a reduced yield.

B This early stage of a fungal infection of wheat leaves is difficult to distinguish from damage caused by weedkiller spray drifting from nearby fields.

Identifying the cause of stress usually begins with careful observation of the plants, looking for visible symptoms. These can include changes in growth, changes in colour or blotching of leaves, or **lesions** (areas of damage) on stems or leaves. Symptoms can then be checked against photos or other information to suggest a cause.

 2 a Describe the visible symptoms of damage on the plants in photo B.

 b Explain why the visible symptoms were not enough to identify the cause of the damage.

Symptoms for different problems may look similar, and symptoms for the same problem may look different in different plants in the same crop. So it can help to use **distribution analysis**, which looks at *where* the damaged plants occur. Flooding, drought or lack of a soil nutrient will create similar symptoms in all the plants in the area. Diseases that spread by wind will affect plants over a wide area, though most obviously where the wind first reaches the crop. Soil pathogens are usually only found in small areas, and so create an obvious pattern of damaged plants.

H

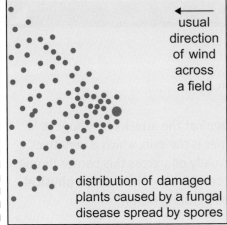

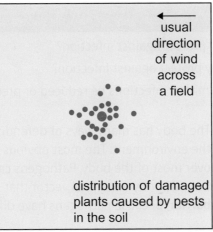

distribution of damaged plants caused by a fungal disease spread by spores

distribution of damaged plants caused by pests in the soil

Key
● first infected plant
∴ plants infected from first infected plant

C If there is a wind, the spread of a plant disease caused by a fungus releasing spores into the wind will form a different pattern of infection compared with a disease caused by insects or by pathogens in the soil.

3 a Potato cyst nematodes are worm pests that eat the roots of potato plants. The nematodes only live in soil. Suggest the distribution pattern of infection that would be found in a field where some potatoes were infected with this pathogen.

b Suggest what the potato grower could do to confirm the presence of nematodes.

Sometimes the only way to get a definite identification of a crop disease is to send samples to a lab for testing. The tests should allow a **diagnosis** of the problem to be made. These tests can include trying to grow a pathogen from damaged crop plants, or using technology to identify the presence of genetic material from a pathogen. When farmers send damaged plants for testing, they will also send in a report about other observations they have made. They may also send soil samples to be tested for nutrients and toxins. This diagnostic testing helps the lab to be more certain of the cause of a problem.

4 a Describe what is meant by diagnostic testing.

b Lab testing for a fungal disease can involve growing the fungus on nutrient agar. Explain why an aseptic technique would be needed for setting up this test.

5 Explain how soil samples and field observations can help the lab make an accurate diagnosis.

D The machine in this photo is used to measure the areas of healthy and diseased leaf, which can distinguish between diseases.

Checkpoint

How confidently can you answer the Progression questions?

Strengthen

S1 A farmer sends a sample of a diseased plant and a sample of the soil it grew in to a lab. Describe one test the lab might carry out on each sample to help find out what caused the disease.

Extend

E1 A farmer finds black spots on the leaves on his crop. Describe how the cause of the black spots could be identified.

Exam-style question

Describe three ways in which plant diseases may be identified. *(3 marks)*

SB5i Physical and chemical barriers

Specification reference: B5.8; B5.12

Progression questions

- How do physical barriers of the body protect against infection?
- How do chemical barriers of the body protect against infection?
- How can the spread of sexually transmitted infections be reduced or prevented?

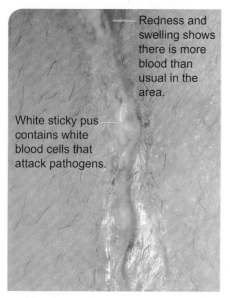

Redness and swelling shows there is more blood than usual in the area.

White sticky pus contains white blood cells that attack pathogens.

A Although blood clotting quickly blocks an open wound, infection by pathogens can cause swelling and the formation of pus as the body attacks the pathogen.

The body has many ways of defending against the attack of pathogens in the environment. The most obvious barrier is the skin, which is very thick over most of the body. Pathogens can usually only cross this barrier through wounds or by an animal vector that pierces the skin. The skin is a **physical barrier**, because pathogens have difficulty getting past it.

 1 Name one protist pathogen that is able to get through the skin barrier, and explain how it does this.

The skin has additional defences, because it contains glands that secrete substances onto the skin surface. These substances include **lysozyme**, which is an enzyme that breaks down the cell walls of some types of bacteria. Lysozyme is a **chemical defence**, because it reacts with substances in the pathogen and this kills the pathogens or makes them inactive.

Did you know?

Alexander Fleming named lysozyme in 1922, when he found mucus from the nose of a person with a cold killed bacterial colonies on an agar plate. Later work on lysozyme identified the way all enzymes work as biological catalysts.

 2 Explain why lysozyme is said to be a chemical defence of the body.

Lysozyme is secreted in tears (from the eyes), saliva (in the mouth), and in **mucus**, where it helps to protect the thinner surfaces of the body. Mucus is a sticky secretion produced by cells lining the many openings, such as the mouth and nose, that pathogens could use to enter the body. Dust and pathogens get trapped in the mucus.

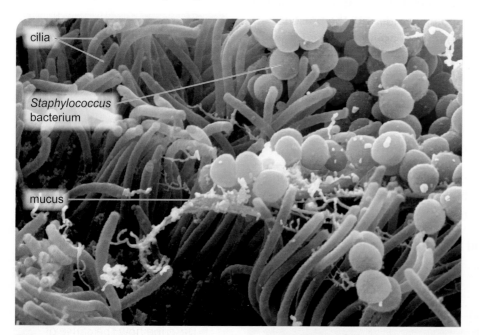

cilia

Staphylococcus bacterium

mucus

B *Staphylococcus* bacteria are trapped in mucus in the nose (magnification × 12 500). The mucus is then wafted by movement of the cilia to the back of the throat, where it can be swallowed.

 3 Is mucus a physical barrier or chemical defence of the body? Explain your answer.

Some of the cells that line the inside of the nose and tubes in the breathing system have cilia. **Ciliated cells** are specialised to move substances such as mucus across their surfaces. This helps to carry dust and pathogens away, either out of the body or into the throat, where they enter the digestive system.

Food, drink and mucus from the respiratory system all pass down the oesophagus (gullet) into the stomach. Some of the cells lining the stomach secrete **hydrochloric acid**, reducing the pH of the stomach contents to about 2. At this acidity, many pathogens are destroyed. Only a few types of bacteria, such as *Helicobacter pylori*, are adapted to survive in the stomach.

 4 Explain how cilia help cells lining tubes in the lungs to carry out their function well.

5 Explain why the stomach is a good defence against pathogens entering the body.

Sexually transmitted infections

The reproductive system has natural defences, such as lysozyme in vaginal fluid and mucus. However, some pathogens can overcome these defences. These pathogens are usually transmitted through sexual activity and are called **sexually transmitted infections** (**STIs**). These include the HIV virus and the *Chlamydia* bacterium. Both of these pathogens can be spread by contact with sexual fluids (semen or vaginal fluid). This method of transmission can be reduced or prevented by avoiding direct contact with sexual fluids, such as by using a condom as an artificial barrier during sexual intercourse. Both of these pathogens may also be passed from a pregnant mother to her unborn baby, which can harm the baby. HIV may also be passed from an infected person to others in blood, such as through sharing needles when injecting drugs.

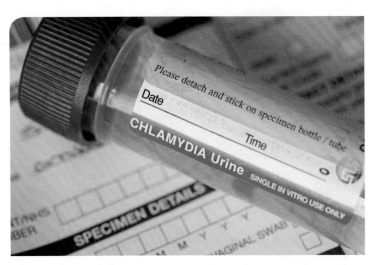

C A small sample of urine or a swab of sexual fluids can be used to screen for *Chlamydia*.

Many people with STIs are not aware that they are infected. **Screening** helps to identify an infection so that people can be treated for the disease. For example, there is a nationwide *Chlamydia* screening programme in the UK for people under 25. As a result, over 200 000 people are being diagnosed with *Chlamydia* each year, and the number of diagnosed new cases is rising. However, this may be due to more people being screened than before.

Blood given to people who have lost a lot of blood, such as during an operation, is first screened (checked) to make sure it does not contain HIV particles or other pathogens.

 6 Explain how screening for an STI can help to reduce spread of the infection.

Checkpoint

How confidently can you answer the Progression questions?

Strengthen

S1 Millions of pathogens are breathed into the nose and mouth every day. Describe all the barriers and defences that the body has to prevent those pathogens causing disease.

Extend

E1 Almost 70 per cent of all *Chlamydia* diagnoses are in people under 25 years old. Suggest reasons for this, and explain your answers.

Exam-style question

Explain how one natural physical barrier and one chemical defence stop pathogens from entering the body.   *(2 marks)*

SB5j The immune system

Specification reference: B5.13; B5.14; B5.15B

Progression questions

- What is the function of the immune system?
- What are the stages of response by the immune system to infection?
- How does immunisation protect the body from disease?

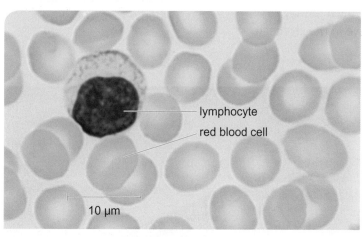

A Lymphocytes are one type of white blood cell. Their function in the immune system is to make antibodies. Different lymphocytes make different antibodies.

Sometimes pathogens manage to get through all the physical barriers and chemical defences of the body. It is then the job of the immune system to attack the pathogens and try to prevent them causing harm.

All cells and virus particles have molecules on their outer surfaces called **antigens**. The immune system uses these antigens to identify if something inside the body is a cell of the body or has come from outside.

1 What is an antigen?

2 Explain why the immune system attacks pathogens but not other cells in the body.

1 Pathogens have antigens on their surface that are unique to them.

These lymphocytes are not activated.

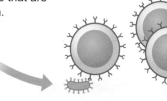

2 A lymphocyte with an antibody that perfectly fits the antigen is activated.

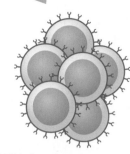

3 This lymphocyte divides over and over again to produce clones of identical lymphocytes.

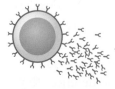

4 Some of the lymphocytes secrete large amounts of antibodies. The antibodies stick to the antigens and destroy the pathogen. Other lymphocytes remain in the blood as memory lymphocytes, ready to respond immediately if the same antigen ever turns up again.

B how the immune system attacks a pathogen

White blood cells called **lymphocytes** have other molecules on their surface, called **antibodies**. A lymphocyte with antibodies on its surface that match the shape of the antigens on a pathogen will attach to the pathogen. This stops the pathogen from working. The lymphocyte is **activated** and will divide rapidly to produce many identical lymphocytes with the same antibodies. Some of these cells release large amounts of identical antibody molecules into the blood. The antibodies attach to pathogens with the matching antigens. This may kill the pathogens or cause other parts of the immune system to destroy them.

3 Explain why only some lymphocytes are activated by a pathogen.

When all the pathogens have been killed, some of the lymphocytes with antibodies that match that pathogen's antigens remain in the blood. These cells are **memory lymphocytes**. If the same kind of pathogen tries to infect you again, the memory lymphocytes cause a much faster **secondary response** that will stop you becoming ill. This means you are **immune** to that pathogen. Immunity to one pathogen does not make you immune to a different pathogen, because different pathogens need different antibodies to attack them.

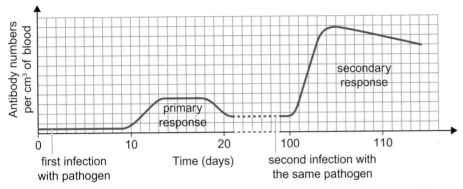

C The immune responses to the first and second infection by a pathogen are different.

Immunisation

Immunity to a pathogen (**immunisation**) can be triggered artificially, by using a **vaccine**. The vaccine contains weakened or inactive pathogens, or bits of the pathogen that include the antigens. The vaccine may be injected into the body, or be taken by mouth, and usually causes little reaction. Most vaccines will protect against particular diseases for many years.

 5 What is a vaccine?

 6 Explain why immunisation against one pathogen does not produce immunity to a different pathogen.

In a very few cases of immunisation, a person may react badly to a vaccine. For example, about 1 in 900 000 doses of **MMR** (measles, mumps, rubella) vaccine cause a high temperature or even fits. If there is a known risk that a child might react badly, then that child may not be given the vaccine. However, they will still be protected if around 95 per cent of other children are immunised, because their chance of coming into contact with an infected person will be very low. This is known as **herd immunity**.

 7 About 1 in 5000 people who have measles die from the disease, and about 1 in 15 will suffer severe reactions to the virus. Describe advantages and disadvantages of immunisation against measles.

8 Explain how herd immunity works.

Exam-style question

Explain why vaccination against measles makes you immune to measles.

(2 marks)

 4 a Use graph C to identify two ways in which the response to the first and second infection by the same pathogen differ.

 b Suggest how the differences in response explain why you may feel ill on the first infection but not the second.

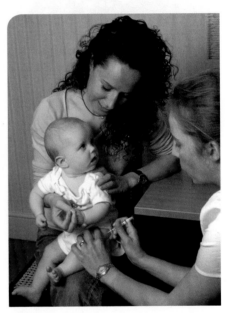

D Many childhood diseases that were once common, such as diphtheria and measles, are now rare in the UK because most children are immunised against the diseases.

Checkpoint

How confidently can you answer the Progression questions?

Strengthen

S1 How does the body respond if a pathogen gets past the body's natural defences?

Extend

E1 Compare the body's natural response to infection with immunisation.

SB5k Antibiotics

Specification reference: B5.16; B5.20

Progression questions

- What are antibiotics?
- Why are antibiotics useful?
- How are new medicines developed?

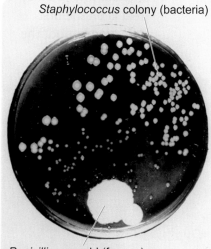

A The bacterial plate that helped Fleming discover penicillin. Each bacterial **colony** contains large numbers of bacteria.

In 1928, Alexander Fleming noticed something strange on an agar plate covered in bacteria that he had left for several weeks. Where a mould had grown, the bacteria had been killed. He had discovered that the mould made **penicillin**.

Further work was needed to extract and purify the penicillin and to make it in large quantities, but it became the first antibiotic. **Antibiotics** are substances that either kill bacteria or **inhibit** their cell processes, which stops them growing or reproducing. Antibiotics do not have this effect on human cells. This makes them useful for attacking bacterial infections that the immune system cannot control.

1 What evidence in photo A suggests the mould is releasing a substance that kills bacteria?

2 Explain why giving penicillin to people with serious wounds could help save their lives.

3 Explain why antibiotics have no effect on diseases such as flu and HIV.

Many kinds of antibiotic have been developed that work in different ways. This is important because different types of bacteria have different structures and they do not all respond in the same way to a particular antibiotic.

Did you know?

Penicillin was first used on a large scale during the Second World War. It saved thousands of soldiers from agonising deaths due to infected wounds.

4 Explain how ceftazidime works as an antibiotic.

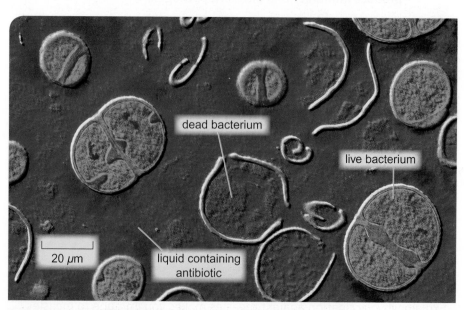

B The antibiotic ceftazidime works by damaging bacteria that have a particular cell wall structure, so that they break open.

A major problem with using antibiotics is that many kinds of bacteria are evolving **resistance**, so they are no longer harmed by the antibiotic. New antibiotics and other medicines must be developed to help control infection.

The first step in the development of a possible new medicine is when it is tested on cells or tissues in the lab. This is the first **pre-clinical** stage of testing. This stage shows if the medicine can get into the cells and have the required effect. Although we take medicines to make us better, all medicines have **side effects**, causing unintended changes that may be harmful. So testing tries to make sure that harmful side effects are limited.

If this first stage is successful, the new medicine may then be tested on animals to see how it works in a whole body (without risk to humans). If that stage is successful, the medicine is tested in a small **clinical trial**, on a small number of healthy people, to check that it is safe and that side effects are small.

If that stage is successful, the medicine is then used in a large clinical trial, on many people who have the disease that the medicine will be used to treat. This helps to work out the correct amount to give (the **dose**), and to check for different side effects in different people. Only if a new medicine passes all these tests can a doctor prescribe it for treating patients.

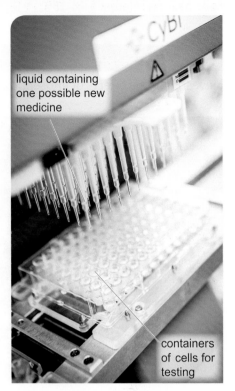

liquid containing one possible new medicine

containers of cells for testing

C Many cell cultures can be automatically tested with several possible new medicines at the same time. This increases the rate of discovery of new medicines.

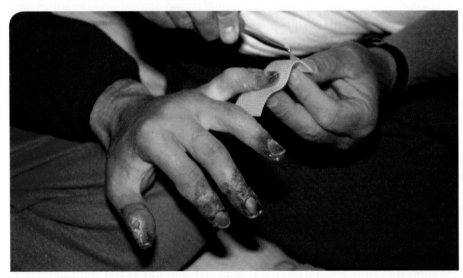

D Clinical trials do not always go according to plan. In one case, six men were injected with a trial leukaemia drug called TGN1412. It caused organ failures and fingers and toes to turn black (some of which had to be amputated).

 5 a List the stages of developing a new medicine, from discovery to prescription.

 b Explain why each stage is needed.

Exam-style question

Describe two stages of pre-clinical testing in the development of a new antibiotic. *(2 marks)*

Checkpoint

How confidently can you answer the Progression questions?

Strengthen

S1 A new antibiotic has been made. Describe how the antibiotic will be tested before doctors are allowed to use it on their patients.

Extend

E1 A new medicine can only move to the next stage of testing when it has been successful in the previous stage. Describe the advantages and disadvantages of this, including time and cost of development.

SB5k Core practical – Antibiotics

Specification reference: B5.18B

Aim

Investigate the effects of antiseptics, antibiotics or plant extracts on microbial cultures.

A Alexander Fleming's untidy way of working led to the discovery of penicillin.

The discovery by Alexander Fleming that *Penicillium* mould produced an antibiotic would not have happened if he had used aseptic techniques. Fleming was studying *Staphylococcus* bacteria, which cause sore throats and skin infections. On returning from holiday, he noticed that one of his bacterial cultures was contaminated with mould. This led to the discovery of penicillin (see topic *SB5k Antibiotics*).

Microbial cultures (for example, of certain bacteria) are used to study the effects of plant extracts (see topic *SB5g Plant defences*), antibiotics and **antiseptics**. (Antiseptics are substances used to kill microorganisms on the surface of the body or on equipment.) In this kind of investigation it is important to work aseptically so that the substances are only tested against one organism and the results are not spoilt by other microorganisms.

Your task

You will prepare bacterial lawn plates, then use discs containing different concentrations of an antibiotic (or antiseptic) to investigate the effect on bacterial growth. It is essential that you use aseptic techniques as you work.

Method

Stage 1

A Use aseptic technique to pour an agar plate. Make sure the base of the Petri dish is covered with agar and the surface of the agar is smooth.

B When the agar has set, remove a sterile pipette from its wrapper. Do not put this down.

C With your left hand pick up the bottle of bacterial culture. Place its cap in the palm of your right hand, and curl the little finger of your right hand around the cap. Then twist the cap off, keeping hold of it.

D Pass the neck of the bottle through a Bunsen flame, insert the pipette into the culture and draw up a very small amount of the culture.

E Pass the neck of the bottle through the Bunsen flame again, and screw the cap back on.

F Lift the lid of the Petri dish a small way, and gently add a couple of drops of culture to the agar. Replace the lid. Place the pipette in disinfectant.

G Unwrap a sterile spreader. Lift the lid of the Petri dish a small way, and then spread the drops of the culture across the agar using a side-to-side motion with the spreader. Replace the lid. Place the spreader in disinfectant.

Stage 2

H Mark the bottom of the dish in sections, as shown in diagram B. Label one section for each concentration of the antibiotic you will use and one for the control. Place the dish the right way up on the bench.

I Sterilise the forceps and use them to place a sterile filter paper disc in the 'control' section of the dish.

J Re-sterilise the forceps, then use them to place one antibiotic disc in the correct section of the dish.

K Repeat step J for the other antibiotic discs.

L Tape the lid onto the dish as shown in diagram C and turn the dish upside down. Give the dish to your teacher for incubation. Wash your hands thoroughly.

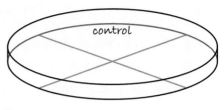

B

Stage 3

M Measure the diameter of the clear space around each disc. Work out the radius and use this to calculate the cross-sectional area of each space.

N Draw a graph showing cross-sectional area against concentration of antibiotic.

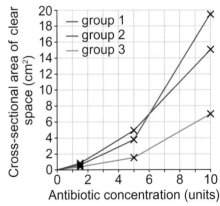

C

Exam-style questions

1 Describe how you would sterilise the forceps and control filter paper disc. *(1 mark)*

2 Explain the aseptic technique used in steps I and J. *(2 marks)*

3 Explain the purpose of the control disc. *(2 marks)*

4 Give a reason why the taped dishes are kept in a warm place. *(1 mark)*

5 Explain why there are clear areas around the antibiotic discs after the dish has been kept warm for a few days. *(2 marks)*

6 Describe how you would obtain measurements for the radius of a clear space accurately. *(1 mark)*

Three groups of students carried out the experiment described in the method. Their results are shown in graph D.

7 a Describe the difference in results between group 3 and the other groups. *(1 mark)*

b Suggest a possible reason for the difference. *(1 mark)*

c Justify your answer to part **b**. *(1 mark)*

8 Draw a conclusion from the results of groups 1 and 2 shown in the graph. *(2 marks)*

In a different experiment, different antibiotics were tested on a culture of the bacterium *Micrococcus luteus* using a similar method. The results are shown in table E.

9 Calculate the cross-sectional area of each clear space. *(3 marks)*

10 Draw a conclusion from the results shown in the table. *(2 marks)*

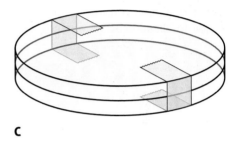

How antibiotic concentration affects area of clear space around disc

D

Antibiotic	Radius of clear disc (cm)
benzylpenicillin	1.4
meticillin	0.05
streptomycin	1.8

E

SB5l Monoclonal antibodies

Specification reference: H B5.21B; H B5.22B

Progression questions

- H What are monoclonal antibodies?
- H How are monoclonal antibodies produced using hybridoma cells?
- H How are monoclonal antibodies used?

H

This line contains antibodies to the hormone that is produced soon after fertilisation of the egg cell.

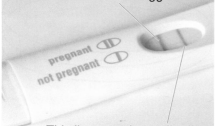

This line contains antibodies to a protein always found in urine and shows that the test has worked properly.

A If the target substances (antigens) attach to the monoclonal antibodies on the pregnancy test stick, a dye turns from colourless to red.

Pregnancy test sticks are a simple way for a woman to tell if she is pregnant using her urine. They work by detecting a hormone that is produced only in pregnancy. The hormone is present in tiny amounts, so the test has to be extremely sensitive. This sensitivity is achieved using antibodies specially made to match the hormone.

Pregnancy test sticks need large amounts of identical antibodies, called **monoclonal antibodies**. Monoclonal antibodies cannot be made in large amounts using normal lymphocytes. Although a lymphocyte can divide over and over again to make many copies of itself (**clones**), once it has started to produce antibodies it cannot divide any more. To get around this problem, **hybridoma cells** are made by fusing a lymphocyte that produces the right kind of antibodies with a **cancer cell**. Cancer cells are used because they can divide over and over again. Hybridoma cells have the characteristics of the lymphocyte *and* the cancer cell they were made from.

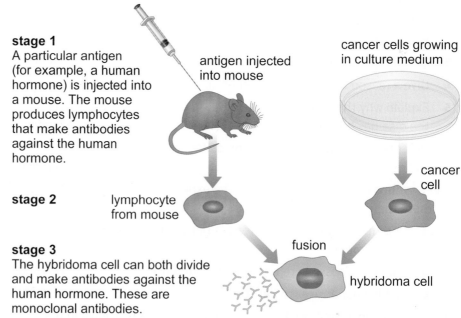

stage 1
A particular antigen (for example, a human hormone) is injected into a mouse. The mouse produces lymphocytes that make antibodies against the human hormone.

antigen injected into mouse

cancer cells growing in culture medium

cancer cell

stage 2

lymphocyte from mouse

fusion

stage 3
The hybridoma cell can both divide and make antibodies against the human hormone. These are monoclonal antibodies.

hybridoma cell

B how monoclonal antibodies are made

Monoclonal antibodies can be made to match and stick to any kind of protein, such as hormones and enzymes. They can be made to match the antigens on pathogens (and so help to identify the pathogens). They can also be made to stick to specific cells in the body such as cancer cells or **platelets**. (Platelets are fragments of blood cells that help to form blood clots. In the wrong places, such as the brain or heart, these clots can kill.)

6th **1** Define the term monoclonal antibodies.

7th **2** Explain why hybridoma cells are needed to produce monoclonal antibodies.

8th **3** Describe how the monoclonal antibodies needed for a pregnancy test stick are made.

Since monoclonal antibodies can be made to stick to certain types of cell, they can be used in medical **diagnosis**. This is often done by making the antibodies slightly radioactive. When the antibodies attach to cancer cells, the radioactivity can be detected (using a **PET scanner**, for example) and so the position of the cancer can be found.

4 Explain how monoclonal antibodies could be used to identify where a blood clot has formed in a patient's brain.

Many people with cancer are given drugs (**chemotherapy**) or ionising radiation (**radiotherapy**) to kill the cancer cells. These treatments also expose healthy cells to the drugs or radiation and can often damage them. Cancer drugs can be attached to monoclonal antibodies so that they are delivered just to the cells that need treating. This reduces the amount of drug needed to kill the cancer cells and reduces the risk of damaging healthy cells.

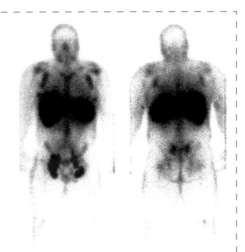

C Radioactive monoclonal antibodies can make it easier to pinpoint cancer cells and tumours (shown in red in this scan)

D Cancer treatment can affect healthy cells in the body, such as those that make hair and new blood cells. The healthy cells usually recover after treatment has ended.

5 Describe how monoclonal antibodies that deliver drugs to kill cancer cells could be produced.

6 Explain why using monoclonal antibodies in the treatment of cancer could be better for the patient than traditional chemotherapy or radiotherapy.

Exam-style question

Describe how monoclonal antibodies can be used to find the position of cancer cells in the body. *(2 marks)*

Checkpoint

How confidently can you answer the Progression questions?

Strengthen

S1 How are monoclonal antibodies used in the diagnosis of some diseases and the treatment of those diseases?

Extend

E1 During the clotting of blood an enzyme changes one substance into another, which then forms the clot. Describe how monoclonal antibodies could be made that bind to the enzyme, and suggest how they could be tested to make sure these monoclonal antibodies are effective in preventing blood clots in humans.

Immunisation

Young children are immunised against a range of infectious diseases. Explain how immunisation protects them from these diseases.

(6 marks)

. . . .

Student answer

Immunisation means making someone immune to a disease by giving them a vaccine [1]. The vaccine contains an inactive form of the pathogen [2] that causes the disease. As these are not active pathogens, it won't make the person suffer from the disease.

The vaccine contains pathogen antigens. Putting these antigens into a child's body causes an immune response, which means that [3] lymphocytes that match these particular antigens become activated and produce many matching lymphocytes and antibodies. Some of the lymphocytes become memory lymphocytes and remain in the blood for a long time.

If the live pathogen gets into the body at a later time, the memory lymphocytes are already there to recognise the antigens on the pathogen and cause an immune response. This response is large and rapid because a large amount of antibodies is produced very quickly. The antibodies attack the pathogens and kill them before they can make the child ill [4].

[1] The student has explained a key scientific term from the question. This is often a good way to introduce an answer.

[2] There is good use of correct scientific terms (e.g. pathogen, vaccine, antigen and lymphocyte) and these have all been spelt correctly, which is important.

[3] The answer contains clear links between ideas by using linking words or phrases, such as 'because' or 'which means that'.

[4] The question has been fully answered and explains how immunised children are protected against infectious diseases.

. . . .

Verdict

This is a strong answer. It explains what immunisation is, and then clearly sets out the steps by which a vaccine causes an immune response, leading to protection from a particular disease. The answer also uses proper scientific terms in a way that makes their meanings clear.

Exam tip

The effect of immunisation has a clear sequence, so it is important to make sure this is ordered in a logical way and linked back to the question being asked. To plan an answer like this, write out key words and phrases and then put them in order by using numbers. Also, write some key linking words and phrases that you will use, then cross these notes out as you include them in your answer.

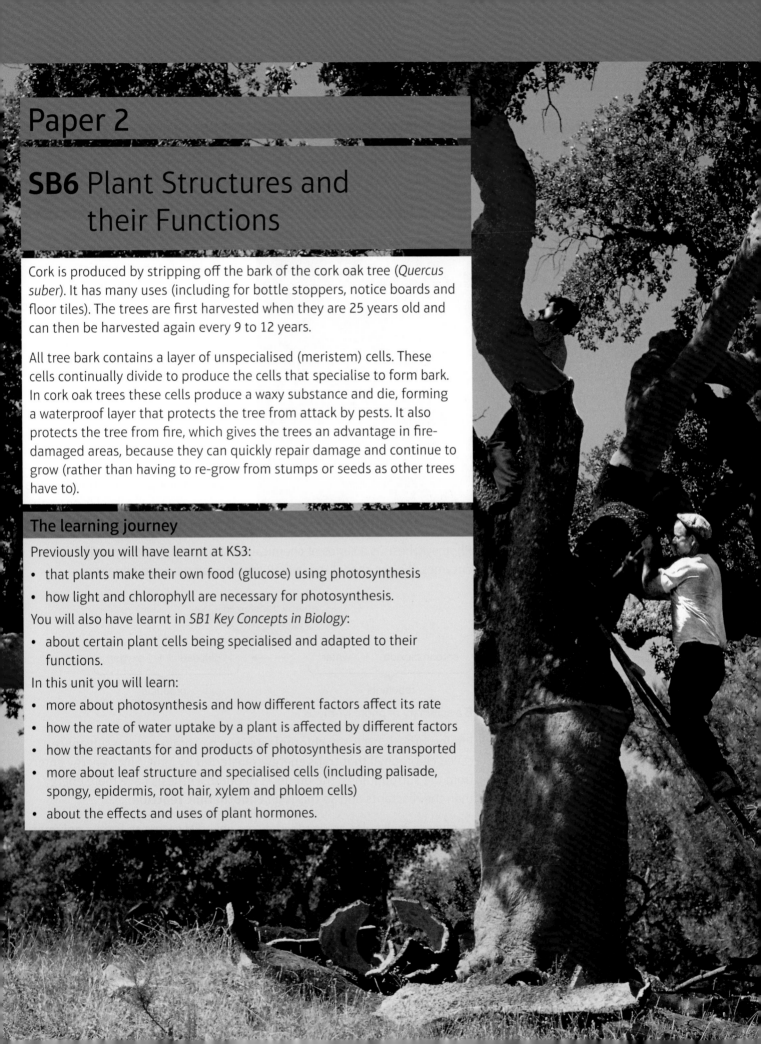

Paper 2

SB6 Plant Structures and their Functions

Cork is produced by stripping off the bark of the cork oak tree (*Quercus suber*). It has many uses (including for bottle stoppers, notice boards and floor tiles). The trees are first harvested when they are 25 years old and can then be harvested again every 9 to 12 years.

All tree bark contains a layer of unspecialised (meristem) cells. These cells continually divide to produce the cells that specialise to form bark. In cork oak trees these cells produce a waxy substance and die, forming a waterproof layer that protects the tree from attack by pests. It also protects the tree from fire, which gives the trees an advantage in fire-damaged areas, because they can quickly repair damage and continue to grow (rather than having to re-grow from stumps or seeds as other trees have to).

The learning journey

Previously you will have learnt at KS3:

- that plants make their own food (glucose) using photosynthesis
- how light and chlorophyll are necessary for photosynthesis.

You will also have learnt in *SB1 Key Concepts in Biology*:

- about certain plant cells being specialised and adapted to their functions.

In this unit you will learn:

- more about photosynthesis and how different factors affect its rate
- how the rate of water uptake by a plant is affected by different factors
- how the reactants for and products of photosynthesis are transported
- more about leaf structure and specialised cells (including palisade, spongy, epidermis, root hair, xylem and phloem cells)
- about the effects and uses of plant hormones.

SB6a Photosynthesis

Specification reference: B6.1; B6.2; B6.9

Progression questions

- What happens during photosynthesis?
- Why is photosynthesis so important for almost all life on Earth?
- How is a leaf adapted for photosynthesis?

A Plant biomass feeds impalas. In turn, impala biomass feeds cheetahs.

All organisms need energy. Plants and algae (a type of **protist**) can trap energy transferred by light from the Sun. This energy is then transferred to molecules of a sugar called **glucose**, in a process called **photosynthesis**. Glucose and substances made from glucose are stores of energy. When animals eat plants, they get the energy from these stores.

The materials in an organism are its **biomass**. Plants and algae produce their own biomass and so produce the food for almost all other life on Earth. They are the **producers** in **food chains**.

5th **1** Explain why most animals depend directly or indirectly on plants and algae.

7th **2** Outline how plants produce biomass.

Photosynthesis is a series of chemical reactions, catalysed (speeded up) by enzymes. We can model the overall process using a word equation.

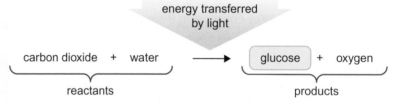

B a summary of photosynthesis

6th **3** What are the reactants in photosynthesis?

9th **4** Explain why the products of photosynthesis have more energy than the reactants.

7th **5** Which product of photosynthesis is needed to make starch?

9th **6** At what time of day would the amount of starch in chloroplasts be at its highest? Explain your reasoning.

Photosynthesis occurs in **chloroplasts**, which contain a green substance called chlorophyll that traps energy transferred by light. Since energy enters from the surroundings, the products of photosynthesis have more energy than the reactants and so this is an **endothermic reaction**.

As glucose molecules are made, they are linked together to form a **polymer** called **starch**. This stays in the chloroplasts until photosynthesis stops. The starch is then broken down into simpler substances, which are moved into the cytoplasm and used to make **sucrose** (another type of sugar molecule). Sucrose is transported around the plant and may be used to make:

- starch (in a **storage organ** such as a potato)
- other molecules for the plant (such as **cellulose**, **lipids** or **proteins**)
- glucose for **respiration** (to release energy).

Leaf adaptations

Leaves are often broad and flat, giving them a large surface area. The **palisade cells** near the top of a leaf are packed with chloroplasts. These adaptations allow a leaf to absorb a great deal of light.

Carbon dioxide for photosynthesis comes from the air. Leaves contain microscopic pores called **stomata** (singular **stoma**). Stomata allow carbon dioxide to diffuse into the leaf. The stomata are opened and closed by specialised **guard cells**. In the light, water flows into pairs of guard cells making them rigid. This opens the stoma. At night, water flows out of the guard cells. They lose their rigidity and the stoma shuts.

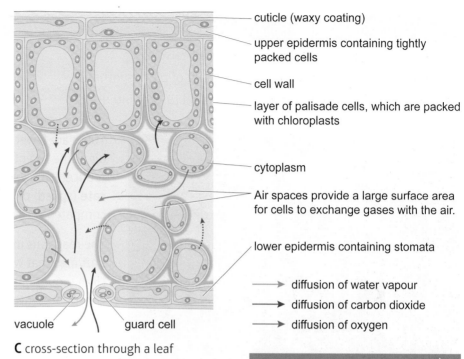

cuticle (waxy coating)

upper epidermis containing tightly packed cells

cell wall

layer of palisade cells, which are packed with chloroplasts

cytoplasm

Air spaces provide a large surface area for cells to exchange gases with the air.

lower epidermis containing stomata

→ diffusion of water vapour

→ diffusion of carbon dioxide

→ diffusion of oxygen

vacuole   guard cell

C cross-section through a leaf

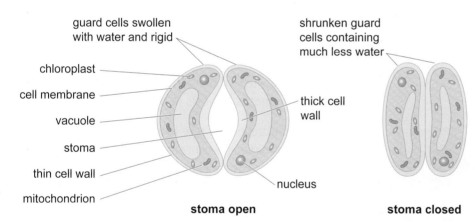

guard cells swollen with water and rigid

chloroplast

cell membrane

vacuole

stoma

thin cell wall

mitochondrion

shrunken guard cells containing much less water

thick cell wall

nucleus

stoma open   **stoma closed**

D Stomata open and shut depending on the amount of light.

Leaves are thin, so carbon dioxide does not have far to diffuse into the leaf before reaching cells that need it. Stomata also allow the oxygen produced by photosynthesis to escape into the air, as well as water vapour. The flow of different gases into and out of a leaf is an example of **gas exchange**.

 7 Explain how stomata open when it is light and close when it is dark.

 8 Water lily leaves float on the surface of ponds and lakes. Suggest how their leaves might be different to the leaf in diagram C. Explain your reasoning.

Exam-style question

Explain how palisade cells are adapted for photosynthesis. *(2 marks)*

Did you know?

At the bottom of the oceans, some microorganisms produce biomass without light. They use the energy transferred by some chemical reactions to drive the production of glucose.

Checkpoint

How confidently can you answer the Progression questions?

Strengthen

S1 Explain why carbon dioxide is vital for your life.

S2 List three adaptations of oak tree leaves for photosynthesis.

Extend

E1 A question on an 'ask a scientist' website reads: 'If plants need energy from light, how do they get their energy when it is dark?' Write an answer to this question.

SB6b Factors that affect photosynthesis

Specification reference: B6.3; **H** B6.4; **H** B6.6

Progression questions

- What are the limiting factors of photosynthesis?
- How do the limiting factors change the rate of photosynthesis?
- **H** How is the rate of photosynthesis related to light intensity?

A Mountain plants, such as edelweiss, are small partly because photosynthesis is slower higher up.

B These plants are being grown in water ('hydroponically') so that water is never a limiting factor. Being in a greenhouse means that the temperature, light intensity and carbon dioxide concentration can be controlled to maximise photosynthesis.

3 Explain why each of the following can be a limiting factor for photosynthesis.

- **a** temperature
- **b** carbon dioxide concentration
- **c** light intensity

4 How would you test the idea that temperature is the limiting factor in graph C?

There are fewer molecules in each cubic centimetre of air at the top of a mountain than at the bottom. This reduced **concentration** of air molecules causes a lower **rate** (speed) of photosynthesis in high mountains compared with sea level.

The reactions in photosynthesis are catalysed by enzymes that work better at warmer temperatures. High mountain areas are cold, which is another reason why photosynthesis is slower at the top of a mountain than at the bottom.

1 What gas from the air is needed for photosynthesis?

2 **a** What is meant by the 'rate' of photosynthesis?

 b State two factors that affect the rate of photosynthesis in mountain areas.

A factor that prevents a rate increasing is a **limiting factor**. Carbon dioxide concentration, temperature and light intensity can all be limiting factors for photosynthesis. The maximum rate of photosynthesis is controlled by the factor in shortest supply.

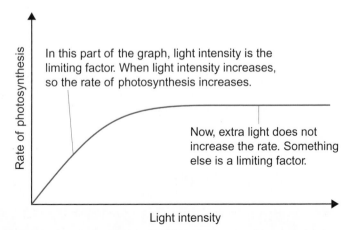

In this part of the graph, light intensity is the limiting factor. When light intensity increases, so the rate of photosynthesis increases.

Now, extra light does not increase the rate. Something else is a limiting factor.

C An increase in light intensity increases the rate of photosynthesis until a limiting factor stops further increases.

126

H

Once a factor is limiting, changing its supply changes the rate of photosynthesis. In graph D, the lower line levels off because temperature is a limiting factor. Increasing the temperature allows higher rates of photosynthesis (upper line).

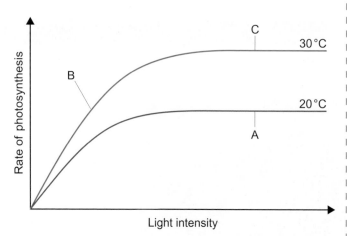

5 In graph C, carbon dioxide concentration is the limiting factor. Sketch a copy of the graph and add another line to predict the effect of increasing carbon dioxide concentration by a set amount.

6 Look at graph D. Explain what factors might be limiting at each stage (A–C).

D The rate of photosynthesis can be increased by increasing the temperature if it is a limiting factor.

A straight line on a graph shows a **linear relationship** between two variables. If the line goes through the origin (0, 0) it shows two variables that are in **direct proportion**. This means that if one variable increases, the other increases by the same percentage. The straight sloping parts of the lines in graphs C and D show that the rate of photosynthesis is directly proportional to light intensity until the limiting factor starts to have an effect.

Inverse square law

To calculate a new light intensity (I_{new}) when the distance of a light source changes (from d_{orig} to d_{new}), we use:

$$I_{new} = \frac{I_{orig} \times d_{orig}^2}{d_{new}^2}$$

I_{new} is **inversely proportional** to d_{new}^2 (light intensity is inversely proportional to the new distance squared). Light intensity varies with distance according to the **inverse square law**.

So, if you double (×2) the distance from a light source, the light intensity is $1/2^2$ or ¼ times the original (it reduces to a quarter of the original). If you halve (÷ 2) the distance from a light source, light intensity is $1/(½)^2 = 1 ÷ ¼ = 4$ times the original.

7 a The light intensity on a plant increases by 3 times and there are no other limiting factors. What is the effect on the rate of photosynthesis? Explain your reasoning.

b How can this increase be achieved by moving the light source?

Checkpoint

How confidently can you answer the Progression questions?

Strengthen

S1 In a forest, some plants grow in the shade and have very large, dark green leaves. Explain this, using the term 'limiting factor' in your answer.

Extend

E1 **H** A light source provides a light intensity of 2000 lux at a distance of 13 cm. Calculate the light intensity when the source is at a distance of 20 cm.

Exam-style question

In a commercial greenhouse, lights switch on at 17:00 in winter. Heating comes on if the inside temperature is below 21 °C. Explain these features.

(2 marks)

Aim

Investigate the effect of light intensity on the rate of photosynthesis.

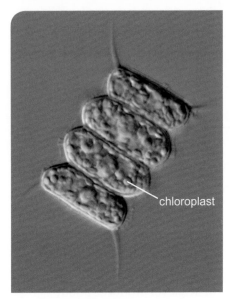

A a small group of *Scenedesmus* algae

Algae are single-celled protists and their cells contain chloroplasts. They photosynthesise in the same way as plants.

Your task

You are going to use balls of jelly containing algal cells to investigate photosynthesis in different light intensities. You can vary the light intensity by altering the distance between a lamp and the algae. You are going to use hydrogen carbonate indicator to monitor the change in pH of the solution in which you placed the balls.

Method

Wear eye protection.

A   Decide on the different distances between the algae and the lamp you are going to use.

B   For each distance you will need one clear glass bottle. You will also need one extra bottle.

C   Add 20 of the algal balls to each bottle.

D   Add the same amount of indicator solution to each bottle, and put on the bottle caps.

E   Your teacher will have a range of bottles showing the colours of the indicator at different pHs. Compare the colour in your tubes with this pH range to work out the pH at the start.

F   Set up a tank of water between the lamp and the area where you will place your tubes. Take extreme care not to spill water near electrical apparatus (such as a lamp).

G   Cover one bottle in kitchen foil, so that it is in the dark.

H   Measure the different distances from the lamp. Place your bottles at those distances. Put the bottle covered in kitchen foil next to the bottle that is closest to the lamp.

I   Turn on the lamp and wait until you can see obvious changes in the colours in your bottles. The longer you can wait, the more obvious your results are likely to be.

J   Compare the colours of all your bottles with the pH range bottles. Write down the pHs of the solutions in your bottles.

K   For each bottle, calculate the 'change in pH/hour'.

L   Plot a suitable graph or chart of your results.

B As the algae photosynthesise, the pH of the solution surrounding them changes. You can add hydrogen carbonate indicator to the solution to detect these changes.

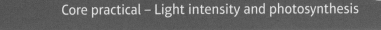

Exam-style questions

1   Explain why the indicator changes colour in the bottles.   *(2 marks)*

2   What is 'change in pH/hour' a measure of?   *(1 mark)*

3   **a**   State the dependent variable in the experiment.   *(1 mark)*

   b   State the independent variable.   *(1 mark)*

   c   One of the control variables is temperature. Explain why this is
a control variable.   *(2 marks)*

   d   Explain whether or not you think this control variable has
been adequately controlled.   *(1 mark)*

   e   List two more variables that have been controlled in this
experiment.   *(2 marks)*

4   The hydrogen carbonate indicator also ensures that there is plenty
of carbon dioxide dissolved in the solution around the algal balls.
Explain the purpose of making sure that there is a good supply of
carbon dioxide.   *(1 mark)*

5   **a**   Describe how you would change this experiment to investigate
the effect of temperature on the rate of photosynthesis.   *(2 marks)*

   b   What temperatures would you choose for your highest and
lowest values? Explain your reasoning.   *(4 marks)*

6   Explain the point of the tube wrapped in kitchen foil.   *(2 marks)*

7   An experiment was carried out using pondweed, as shown in
diagram C. When illuminated, the pondweed produced bubbles
of gas, which were counted as they rose up the inverted funnel.
The number of bubbles was counted for one minute at different
distances from a lamp. Table D shows the results of the experiment.

   a   Name the gas in the bubbles.   *(1 mark)*

   b   Plot the results on a scatter graph and draw a curve of best fit.
        (3 marks)

   c   Explain the shape of your graph.   *(2 marks)*

   d   Explain one way in which this experiment could be improved.
        (2 marks)

   e   **H** Plot a scatter graph to show that the rate of photosynthesis
is proportional to the inverse of the distance squared.   *(4 marks)*

   f   **H** Explain why there is this relationship between the rate of
photosynthesis and the distance in this experiment.   *(2 marks)*

Distance (cm)	Rate of photosynthesis (bubbles per minute)
10	100
15	60
20	30
30	10
40	6
50	4

D

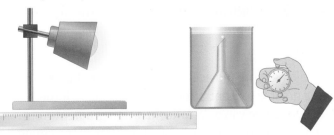

C measuring the rate of photosynthesis in pondweed

Specification reference: B1.15; B6.7

Progression questions

- How are diffusion and osmosis different?
- How do plant roots use diffusion, osmosis and active transport?
- How are root hair cells adapted to their functions?

A This giant buttress-rooted tree grows in Ghana.

Some trees are adapted to living in rainforests by having huge buttress roots. Like all roots, they absorb water and dissolved mineral ions from the soil. However, buttress roots help stop the tall trees falling over in thin rainforest soils, by acting as props. They also trap leaves and other dead vegetation, which then rot to provide additional minerals for the tree.

The water absorbed by plant roots is used for:

- carrying dissolved mineral ions
- keeping cells rigid (otherwise the plants **wilt** – their leaves and stems droop)
- cooling the leaves (when it evaporates from them)
- photosynthesis.

 1 Describe how plants lose water.

 2 For what chemical process do plants need water?

Root hair cells

The outer surfaces of many roots are covered with root hair cells. The 'hairs' are extensions of the cell that provide a large surface area so that water and **mineral ions** can be quickly absorbed. The 'hairs' also have thin cell walls so that the flow of water into the cells is not slowed down.

 3 Suggest two ways in which roots are adapted to absorb a lot of water.

Diffusion and osmosis

A certain volume containing more molecules of a substance than another identical volume has a greater concentration of the molecules. If the two spaces are connected, there will be a **concentration gradient** from higher concentration to lower concentration (see topic *SB1i Transporting substances*).

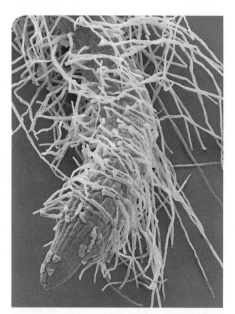

B root hair cells on a newly germinated poppy seedling root (magnification x66)

Particles constantly move in random directions and so particles in a **fluid** spread *down* a concentration gradient. This is **diffusion**. Inside plant roots, the cell walls have an open structure allowing water particles to diffuse towards the middle of the root (from where there are more of them to where there are fewer).

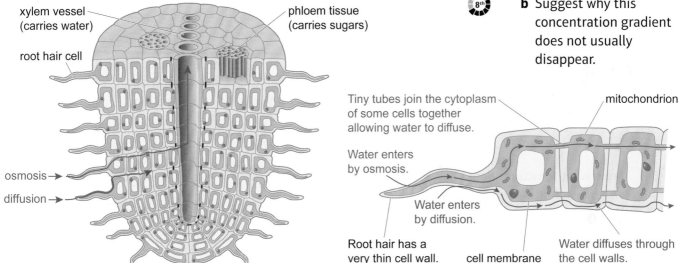

C pathways that water can take through a plant root

Osmosis is when solvent molecules (such as water) diffuse through a **semi-permeable membrane**. They diffuse from where there are more of them (a dilute solution of solutes) to where there are fewer (a more concentrated solution). Cell membranes are semi-permeable and so water passes into the cytoplasm of **root hair cells** by osmosis.

 6 Near the centre of a root is a layer of cells that have a waxy strip running through their cell walls (shown in black on diagram C). This strip stops the diffusion of water. Describe how water diffusing through cell walls gets to the xylem.

Active transport

Mineral salts are naturally occurring ionic compounds. Plants need ions from these compounds to produce new substances. For example, **nitrate** ions are needed to make **proteins**.

The concentration of ions inside a root hair cell is greater than in the soil. Mineral ions cannot diffuse against this concentration gradient. So, proteins in the cell membrane pump the ions into the cell. This is an example of **active transport** (see topic *SB1i Transporting substances*).

 7 Explain why mineral ions do not diffuse into root hair cells.

Exam-style question

Explain how water flows from soil into the cytoplasm of a root hair cell.

(2 marks)

 4 **a** Describe the concentration gradient for water molecules in root cell walls.

b Suggest why this concentration gradient does not usually disappear.

5 By what process does water:

 a enter the cytoplasm of root hair cells

b flow through root cells towards the xylem?

Checkpoint

How confidently can you answer the Progression questions?

Strengthen

S1 Explain how water enters a root hair cell by osmosis.

Extend

E1 Deciduous trees lose their leaves in winter. In the spring, before their leaves emerge, their roots suddenly start to grow and produce new root hair cells. Explain why this happens.

Progression questions

- How do different factors affect the rate of transpiration?
- How is sucrose translocated around a plant?
- How are xylem and phloem adapted to their functions?

Did you know?

A large oak tree loses about 700 litres of water on a hot, sunny day.

 1 From what structures does water evaporate from leaves?

 2 Explain why a tree does not lose much water at night.

 3 Why should leaves be kept cool on a hot day? (*Hint:* think about enzymes.)

 4 State two ways in which transpiration helps a plant.

The evaporation of water from leaves keeps them cool and helps move water (and dissolved mineral ions) up the plant. The flow of water into a root, up the stem and out of the leaves is called **transpiration**.

Xylem vessels form tiny continuous pipes leading from a plant's roots up into its leaves. Inside the vessels is an unbroken chain of water, due to the weak forces of attraction between water molecules. Water is pulled up the xylem vessels in the stem as water evaporates from the xylem vessels in the leaves. As the water vapour diffuses out of a leaf, more water evaporates from the xylem inside the leaf.

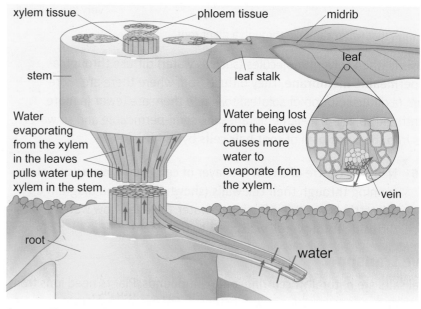

A water flow in a plant

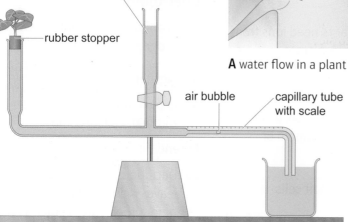

B We can investigate the factors affecting transpiration using a **potometer**. The air bubble moves along the tube as the plant loses water. The speed of the bubble gives a measure of the rate of transpiration (e.g. in mm/min).

The concentration of water vapour in the air spaces inside a leaf is greater than outside it. So, water molecules diffuse down the concentration gradient, out of the leaf. A bigger difference between the concentrations makes the gradient steeper, which makes diffusion faster. So, any factor that reduces the concentration of water molecules outside the stomata will increase transpiration. Factors include:

- wind – moves water molecules away from stomata
- low humidity (little water vapour in the air).

Other factors that increase transpiration are:

- higher temperatures – particles move faster and so diffuse faster
- greater light intensity – makes the stomata wider.

 5 a Explain why water vapour diffuses more quickly out of a leaf on a windy day than on a calm day.

 b Explain why water moves out of the xylem inside leaves during the day.

 6 Explain why, during summer, trees lose more water at midday than at the end of the day.

Xylem

During their development, xylem cells die and their top and bottom cell walls disintegrate. This creates long empty vessels (tubes) through which water can move easily. Xylem vessels are rigid because they have thick side walls and rings of hard **lignin**, and so water pressure inside the vessels does not burst or collapse them. The rigid xylem vessels also help to support the plants.

 7 Describe how xylem cells are adapted to allow water to flow through them easily.

Phloem

Plants make sucrose from the glucose and starch made by photosynthesis. Sucrose is **translocated** (transported) in the **sieve tubes** of the **phloem tissue**. The large central channel in each sieve cell is connected to its neighbours by holes, through which sucrose solution flows.

Companion cells actively pump sucrose into or out of the sieve cells that form the sieve tubes. As sucrose is pumped into sieve tubes (e.g. in a leaf), the increased pressure causes the sucrose solution to flow up to growing shoots or down to storage organs.

 8 Explain why sieve cells have little cytoplasm.

 9 Explain why companion cells in a leaf contain so many mitochondria.

Thick side walls and rings of lignin form rigid tubes that will not burst or collapse, and that provide support.

The dead cells have no cytoplasm and so form an empty tube for water to flow through.

one cell

Tiny pores allow water and mineral ions to enter and leave the xylem vessels.

The lack of cell walls between the cells means that water flow is not slowed down.

C xylem adaptations

Holes in the ends of the cell walls allow liquids to flow from one sieve cell to the next.

pore through which sucrose solution can be pumped

mitochondrion

vacuole

The very small amount of cytoplasm (and no nucleus) means that there is more room for the central channel.

companion cell

sieve cell

D phloem adaptations

Checkpoint

How confidently can you answer the Progression questions?

Strengthen

S1 Design a table to compare transpiration and translocation.

Extend

E1 Explain why water enters sieve tubes from neighbouring xylem vessels as sucrose is pumped into the sieve tubes.

Exam-style question

In an experiment with a potometer, a faster wind speed causes the air bubble to move more quickly than a slower wind speed. Explain this observation.

(2 marks)

SB6e Plant adaptations

Specification reference: B6.11B; B6.14B

Progression questions

- How is the structure of a leaf adapted for photosynthesis and gas exchange?
- Why do some plants have needle-shaped leaves?
- How do plants reduce water loss?

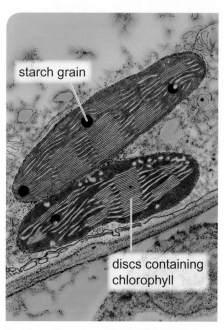

A chloroplasts in a pea leaf (×17 000)

Leaves often have large surface areas to ensure they collect enough light. The energy transferred by light is trapped by chlorophyll packed into discs inside chloroplasts. This energy is then transferred to glucose during photosynthesis. The chloroplasts in a cell can move towards light or away from it (as protection from damage by very bright light).

Carbon dioxide for photosynthesis diffuses into the leaf through open stomata. Leaves are thin so that carbon dioxide does not have far to diffuse before reaching photosynthesising cells. The irregularly shaped **spongy cells** do not fit together well and create air spaces, allowing gases to diffuse easily inside a leaf. A network of xylem vessels supplies the water for photosynthesis.

 1 Explain how you would expect chloroplasts to move as the Sun sets on a sunny day.

 2 Explain why there are starch grains in photo A.

Did you know?

In 1967, Lynn Margulis (1938–2011) suggested that, millions of years ago, photosynthesising bacteria started living inside cells that could not photosynthesise. These bacteria evolved to become chloroplasts in the first plant cells. This is the accepted theory today.

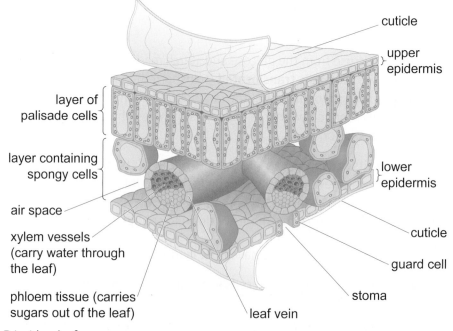

B inside a leaf

 3 Look at diagram B. Where does most photosynthesis occur? How can you tell?

 4 Explain how the shape of the spongy cells helps photosynthesis.

Epidermis cells form the outer layers of a leaf, holding the leaf together and protecting the cells inside. Epidermis cells are transparent, allowing light to pass through them easily. They also produce a waterproof waxy **cuticle**, which helps to prevent water loss. The cuticle also helps to stop microorganisms and water entering the leaf.

5 Explain how the transparent epidermal cells help the function of the leaf.

 6 Suggest an advantage of plants having a cuticle that keeps their leaves clean.

Adaptations for extremes

Plants growing in cold, dry, hot or windy places have adaptations to help stop water loss.

In winter, many broad-leaved **deciduous** plants lose all their leaves, preventing water loss when soil water may be frozen. However, most **conifers** (e.g. pine trees) do not do this. Conifers have needle-shaped leaves with a much smaller surface area and a very thick cuticle. Additionally, this shape creates less wind resistance than broad leaves, allowing conifers to withstand high winds. It also means that they collect less snow.

Plants can reduce water loss by trapping water vapour close to their leaves, which slows the rate of diffusion out of the leaves. Conifers achieve this by having **stomata** located in small pits, where water vapour collects because it is less exposed to air movement. Other plants use tiny hairs to trap water vapour.

Desert habitats are very difficult environments for plants to survive. Photo D shows some adaptations of cacti to living in very dry conditions.

C The cuticle of some plants is so good at repelling water that the leaves become self-cleaning, such as the leaves of the lotus plant. Chemists are investigating leaf cuticle compounds to help develop new water-repellent materials.

 7 Suggest an explanation for why plants underneath the trees in a rainforest often have big leaves.

 8 Explain how the adaptations of their leaves help conifers to conserve water.

Some species have 'hairs'.

In some species, the stomata are only found in the 'valleys' in the stem.

Thick cuticle

Spines, instead of leaves, minimise the surface area of the plant (and protect the stem from herbivores).

Cacti stomata only open at night. Carbon dioxide is taken in at night and stored for use during the day.

stem stores water

D Cacti are adapted to dry environments.

 9 Explain why cactus stomata open only at night.

 10 Small mountain plants often grow in dense groups. Suggest how this may reduce water loss.

Checkpoint

How confidently can you answer the Progression questions?

Strengthen

S1 Draw a table to explain the adaptations that allow cacti to grow in dry deserts.

Extend

E1 Marram grass lives on sand dunes, which dry out quickly. Stomata are found on the surface of the leaf inside pits. Hairs grow over these pits. When the environment is very dry, the leaves curl into tubes (with the pits on the inside). Explain the function of each of these features of the plant.

Exam-style question

Manzanita bushes live in deserts. They have small leaves that have a very thick cuticle. Explain the functions of these features of the plant. *(3 marks)*

SB6f Plant hormones

Specification reference: B6.15B

Progression questions

- What are the names of some plant hormones?
- What are positive and negative phototropism and gravitropism?
- How do auxins cause tropisms in shoots and roots?

A a plant tropism

In biology, a **stimulus** is a change in the environment that causes a **response** by an organism. For example, the plant in photo A has been knocked over, causing a change in the direction of light and gravity acting on the stem. In response, the plant has grown a bend in the stem to make it upright again.

1 Suggest why it is an advantage for the plant in photo A to grow upright.

2 Identify the stimulus and the response shown in photo A.

3 Name the tropism due to light shown in photo A. Include its direction.

4 Describe the effect negative phototropism has on the growth of plant roots.

5 Woodlice move away from light. Explain why this is *not* an example of negative phototropism.

Phototropism

Responding to a stimulus by *growing* towards or away from it is called a **tropism**. A tropism caused by light is a **phototropism**. A tropism *towards* a stimulus is a positive tropism. Plant shoots are positively phototropic, so the plant gets enough light for photosynthesis. Plant roots are negatively phototropic.

Plants respond to stimuli using **plant hormones**. Positive phototropism is caused by plant hormones called **auxins**.

Auxins are produced in the tips of a shoot, where they cause elongation of the cells. If a shoot is grown with light coming from only one direction, auxins move to the shaded side of the shoot. This makes the cells on the shaded side elongate more, which in turn causes the shoot to grow towards the light.

Did you know?

In the Middle Ages, plants that had bends in them (to grow towards light) were given as a treatment for snake bites because they look like snakes.

6 **a** What effect do auxins have on the cells in shoot tips?

b Explain why this is useful for the plant.

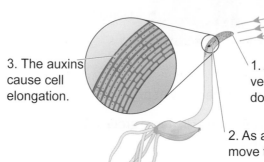

3. The auxins cause cell elongation.

light

1. Auxins are produced at the very tip of the shoot and move down the shoot.

2. As auxins move downwards, they move to the shaded parts of the shoot.

B Auxins control the growth of plant shoots towards light.

One of the first scientists to try to work out what was going on when plant shoots bend towards the light was Charles Darwin (1809–1882). His experiments, and those of other scientists after him, allowed Frits Warmolt Went (1903–1990) to suggest how auxins work. Went's conclusion has now been supported by a lot of evidence, including the extraction and purification of auxins.

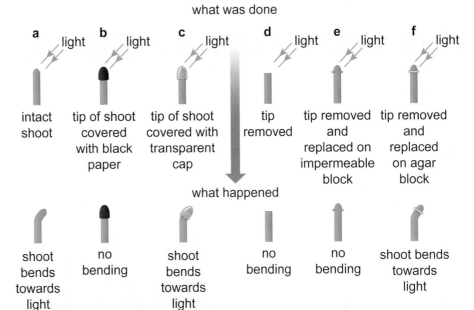

what was done

a intact shoot · **b** tip of shoot covered with black paper · **c** tip of shoot covered with transparent cap · **d** tip removed · **e** tip removed and replaced on impermeable block · **f** tip removed and replaced on agar block

what happened

shoot bends towards light · no bending · shoot bends towards light · no bending · no bending · shoot bends towards light

Darwin's experiments · later experiments

C experiments with phototropism

Gravitropism

Auxins are also found in root tips, where they have the *opposite* effect to that in shoots. In roots, auxins cause cells to stop elongating and this causes positive **gravitropism** – growth towards the direction of gravity. It helps roots to anchor the plant in place and to reach moisture underground.

shoot grows upwards

Auxins are pulled downwards by gravity and inhibit cell elongation.

Auxins are pulled downwards by gravity and increase cell elongation.

root grows downwards

D Auxins have the opposite effects in shoots and roots.

Other plant hormones

Plant hormones called **gibberellins** help seeds to germinate and start to grow roots and shoots. **Ethene** gas is a plant hormone that helps fruit to ripen.

 8 Explain two reasons why a newly germinated root on the surface of the soil will grow downwards.

 9 Identify the roles of the different plant hormones in the life cycle of an apple tree.

 7 Explain why each of the shoots (a–f) in the experiments shown in diagram C has or has not bent towards the light source.

Exam-style question

Explain how auxins cause plant shoots to grow towards a source of light.

(2 marks)

SB6g Uses of plant hormones

Specification reference: **H** B6.16B

Progression questions

- **H** How are auxins used by plant growers?
- **H** What are the uses of gibberellins?
- **H** How do farmers ripen fruit once it has been removed from the tree?

H

Artificial auxins make some plants grow uncontrollably, which can kill them. **Selective weedkillers** contain artificial auxins to kill plants with broad leaves (e.g. dandelions, chickweed) but not those with narrow leaves (e.g. wheat, grass). Farmers can therefore kill weeds in a wheat field without affecting the crop.

A One sprayer nozzle was blocked when a selective weedkiller was sprayed on this wheat field, resulting in a strip of weeds growing amongst the crop.

 1 In what way are artificial auxin weedkillers 'selective'?

 2 Why might a gardener use a selective weedkiller on a lawn?

Auxins are also found in **rooting powders**. The auxins cause plant cuttings to develop roots quickly. Large numbers of identical plants can be produced quickly using cuttings, compared to growing plants from seed.

Gibberellins

Plant hormones called **gibberellins** are naturally released inside a seed to start germination. Some seeds need a period of darkness or cold before they can germinate but plant growers can use gibberellins to make these seeds germinate without this.

Photoperiodism is the response of an organism to the number of daylight hours in a day. Some plants use this to flower at a certain time (e.g. when suitable pollinators are around, or when it is not too cold). Flower growers can override photoperiodism by spraying the plants with gibberellins.

Many plants only produce seeds after being pollinated, which then allows egg cells to be fertilised and seeds to form. Gibberellins can cause some plants to produce fruit without this, so giving us seedless fruits. Gibberellins can also be sprayed on some plants to make them produce bigger fruits.

B a tomato stem a week after being cut and dipped in rooting powder containing auxins

 3 Cuttings can develop roots without rooting powder. What is the advantage of using rooting powder?

H

4   Five seeds of the same type were placed in different strengths of a gibberellin solution (3 mg/dm³, 2 mg/dm³, 1 mg/dm³, 0.5 mg/dm³, 0 mg/dm³) and their germination was monitored.

6th   **a** Which solution was the control?

6th   **b** Explain why a control is needed.

6th   **c** Explain one way in which this experiment could be improved.

8th   **d** Suggest an explanation for why the seed placed in the strongest solution germinated first.

8th   5   Delphiniums only produce their attractive blue flowers during the long days of summer. How could a flower grower get delphiniums to bloom in spring?

8th   6   Gibberellins can be used to make many thousands of seeds all germinate at the same time. Suggest an advantage of this for a garden centre.

Ethene

Fruits that are not ripe (particularly soft fruits, such as bananas) are easier than ripe fruit to transport without damaging them. The unripe fruit can also be kept for longer in cold storage without going off. Fruit producers often pick unripe fruit and then ripen it when needed using a plant hormone gas called **ethene** (or ethylene). This makes sure that fruit reaches the shops in a 'just-ripened' condition.

9th   7   There is a huge range of exotic fruit in supermarkets. Thirty years ago the range was much smaller. Suggest an explanation for this observation.

9th   8   Supermarkets sell 'just ripened' apples all year round, although apples are traditionally harvested in autumn. Explain how this is possible.

D Unripe green bananas from the Caribbean are ripened in the UK using ethene.

Checkpoint

How confidently can you answer the Progression questions?

Strengthen

S1 List the uses of three different plant hormones.

Extend

E1 A factory that tins peaches needs a large amount of peaches that are perfectly ripe all at the same time. Fruits with smaller stones (seeds) also reduce the processing costs of the fruit. A farmer decides to take cuttings from a peach tree that naturally produces fruits containing small stones. He plans to use plant hormones to produce enough peaches that are ready for the factory. Explain how the farmer would use plant hormones.

Mineral ions

Mineral ions are needed by plants in small quantities in order to make certain substances. For example, magnesium ions are an important part of chlorophyll molecules. Explain how magnesium ions are taken into a plant and transported to where they are needed.

(6 marks)

Student answer

The soil contains a lower concentration of magnesium ions than the cells in the roots [1]. Root hair cells pump the magnesium ions into the roots against this concentration gradient using active transport [2]. The ions dissolve in the water in the root cells and are carried along in the water by diffusion [3] into the xylem. The water evaporating from the leaves pulls more water up the xylem [4] and so carries the magnesium ions up the plant to the leaves where it is needed.

[1] A good place to start – at the beginning of the process in the soil.

[2] An excellent point that clearly states why active transport is needed.

[3] There is good use of scientific terms, such as diffusion, active transport and xylem.

[4] There is a clear demonstration that the student understands how the processes of water and mineral ion movement work.

Verdict

This is a strong answer. It has a logically ordered set of sentences that explain the processes that are occurring and makes good use of scientific terms.

Exam tip

While planning your answer, make a list of the important scientific words that you will use. Cross out your list when you have finished writing your answer.

Paper 2

SB7 Animal Coordination, Control and Homeostasis

When a child is hurt, parents and carers will often 'kiss it better'. Many people think this just helps the child to feel better. However, research has shown that this action releases a substance called oxytocin in the person receiving the kiss.

Oxytocin is an example of a hormone and has many effects in the body. It is sometimes called the 'cuddle' hormone because it is released when we are close to people who we are happy being with. One of its other effects, though, is to increase the speed of healing.

So 'kissing it better' might really work!

The learning journey

Previously you will have learnt at KS3:

- about the structure and function of human reproductive systems
- about the menstrual cycle.

You will also have learnt in *SB1 Key Concepts in Biology*:

- how enzymes help digest food molecules.

In this unit you will learn:

- about endocrine glands
- how hormones are transported to their target organs
- how the menstrual cycle is controlled by hormones and how hormones are used in contraception
- about the importance of homeostasis
- about how thermoregulation occurs
- about diabetes and how blood glucose concentration is controlled
- how the kidneys produce urine, and about treatments for kidney failure
- **H** how the hormones thyroxine and adrenalin affect the body
- **H** what a negative feedback mechanism is.

SB7a Hormones

Specification reference: B7.1

Progression questions

- What are hormones?
- Where are hormones produced?
- What are the names of some target organs?

A Response to fear involves the fast responses of the nervous system and the slower responses of the hormonal system, including dilated pupils in the eyes and a faster heart rate.

Your nervous system enables you to respond quickly to changes in your surroundings. However, humans have another response system, called the **hormonal system**. It works more slowly than the nervous system but can cause responses in many parts of the body.

The hormonal system uses chemical messengers called **hormones**, which are carried by the blood and so take time to get around the body. Different hormones are released by a range of **endocrine glands**, including the **pituitary**, **thyroid**, **adrenals**, **ovaries**, **testes** and **pancreas**.

1 The dog in photo A is responding to a threat.

   a Describe as fully as you can how the nervous system has helped the dog identify and respond to the threat.

   b Describe the responses caused by the dog's hormonal system.

   c Compare the speed of response of the nervous and hormonal systems.

2 Define the term 'hormone'.

3 Name one hormone produced in:

   a the ovaries

   b the pancreas.

The pituitary gland releases many hormones, including ACTH, FSH, LH and growth hormone.

The thyroid gland produces several hormones, including thyroxine.

The adrenal glands release several hormones, including adrenalin.

The pancreas contains some cells that produce insulin and others that produce glucagon.

The ovaries produce the sex hormones oestrogen and progesterone.

The testes release the sex hormone testosterone.

B The hormonal system consists of endocrine glands that produce and release hormones.

An organ that is affected by a specific hormone is called its **target organ**. Organs in different parts of the body may be target organs of the same hormone. The hormone affects the organ by changing what the organ is doing. For example, growth hormone stimulates cells in muscles and bones to divide. It also stimulates the digestive system to absorb calcium ions (used to help make strong bones).

C In 2014, Sultan Kösen, the world's tallest man, met Chandra Dangi, the world's shortest man. Kösen was 2.51 m tall while Dangi was 0.55 m. Both had problems with the production of growth hormone in their bodies.

Some endocrine glands are the target organs for other hormones. For example, the **sex hormones** oestrogen and testosterone, which are released by reproductive organs, stimulate the release of growth hormone. The release of sex hormones increases during puberty, which helps to explain the increase in growth rate at this time.

 5 Describe how a change in the amount of sex hormones produced during puberty leads to an increase in growth. Include the names of endocrine glands and target organs for the hormones you mention.

4
  a Name the endocrine gland that produces growth hormone.

  b Name two target organs of growth hormone.

  c Describe how growth hormone reaches its target organs from where it is produced.

  d Suggest why Sultan Kösen is so tall and Chandra Dangi so short.

  e Use your knowledge of how growth hormone works to explain your answer to part **d**.

Checkpoint

How confidently can you answer the Progression questions?

Strengthen

S1 Using a hormone of your choice, state where it is made in the body and where it has its effect.

Extend

E1 Using examples, write definitions of the following terms for a web dictionary: hormone, endocrine gland, target organ.

Exam-style question

Describe how endocrine glands communicate with organs around the body.

(2 marks)

SB7b Hormonal control of metabolic rate

Specification reference: H B7.2; H B7.3

Progression questions

- H What is a negative feedback mechanism?
- H How does thyroxine affect metabolic rate?
- H How does adrenalin prepare the body for 'fight or flight'?

H

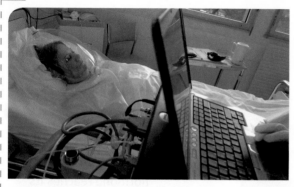

A measuring resting metabolic rate

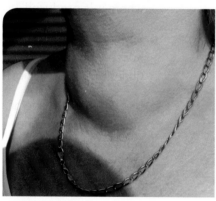

B A goitre is an enlarged thyroid gland. It may be caused by disease or by deficiency of iodine in the diet, because iodine is needed to make thyroxine.

Your **metabolic rate** is the rate at which the energy stored in your food is transferred by all the reactions that take place in your body to keep you alive. **Resting metabolic rate** is measured with the body at rest, in a warm room and long after the person last had a meal.

 1 a Name two processes that require the transfer of energy when the body is fully at rest.

 b Explain why these processes need a source of energy.

 2 Explain why resting metabolic rate is measured in a warm room and not in a cold room.

One hormone that affects metabolic rate is **thyroxine**, which is released by the thyroid gland. Thyroxine is taken into, and affects, many different kinds of cell. It causes heart cells to contract more rapidly and strongly, and it also increases the rate at which proteins and carbohydrates are broken down inside cells.

The amount of thyroxine produced by the thyroid gland is controlled by hormones released by two other glands, as shown in diagram C. The control of thyroxine concentration in the blood is an example of **negative feedback**. This is because an *increase* in thyroxine concentration directly causes changes that bring about a *decrease* in the amount of thyroxine released into the blood, and vice versa.

 3 Suggest an effect on the body caused by the thyroid producing too little thyroxine.

 4 a Define the term 'negative feedback'.

 b Explain why negative feedback is important in controlling thyroxine release.

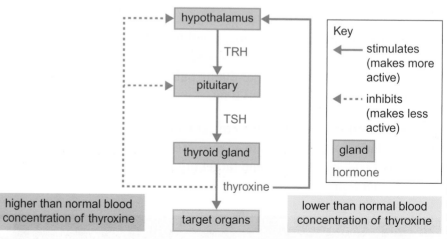

C Negative feedback of the blood concentration of thyroxine involves two other hormones. TRH is thyrotropin-releasing hormone, and TSH is thyroid-stimulating hormone.

H

Adrenalin is a hormone that is released from the adrenal glands. In normal conditions, very little adrenalin is released into the blood. However, in frightening or exciting situations, an increase in impulses from neurones reaching the adrenal glands from the spinal cord triggers the release of large amounts of adrenalin into the blood.

Adrenalin has many target organs, including the liver in which it causes the breakdown of a storage substance called **glycogen**. Glycogen is a polymer made of glucose molecules. When glycogen is broken down, the glucose molecules can be released into the blood providing additional glucose for respiration.

Some of the other target organs and effects of adrenalin are shown in diagram D. Together, these effects prepare the body to fight or run away from danger (the so-called '**fight-or-flight' response**).

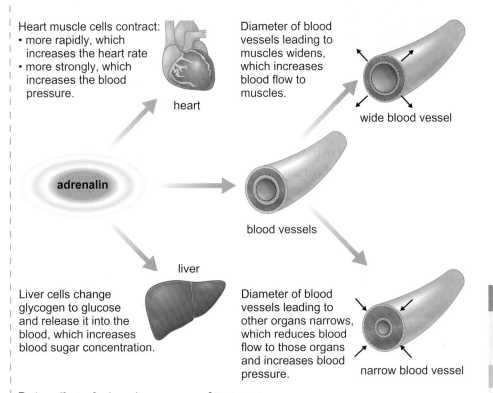

Heart muscle cells contract:
• more rapidly, which increases the heart rate
• more strongly, which increases the blood pressure.

heart

Diameter of blood vessels leading to muscles widens, which increases blood flow to muscles.

wide blood vessel

adrenalin

blood vessels

liver

Liver cells change glycogen to glucose and release it into the blood, which increases blood sugar concentration.

Diameter of blood vessels leading to other organs narrows, which reduces blood flow to those organs and increases blood pressure.

narrow blood vessel

D the effect of adrenalin on some of its target organs

5 a Name three target organs of adrenalin.

   b Explain how the effects of adrenalin on different parts of the body help prepare the body for action.

Checkpoint

How confidently can you answer the Progression questions?

Strengthen

S1 Identify the endocrine gland and target organ(s) for each hormone mentioned on these two pages.

Extend

E1 Compare and contrast how thyroxine and adrenalin are released into the blood, and suggest an explanation for any differences.

Exam-style question

Describe how negative feedback can control the amount of a hormone in the blood. *(2 marks)*

SB7d Hormones and the menstrual cycle

Specification reference: H B7.5; B7.6; H B7.8

Progression questions

- H How do hormones control the menstrual cycle?
- H How do hormones in contraceptive pills interact with hormones in the body to prevent pregnancy?
- H How can hormones increase the chance of pregnancy?

H

A The most babies born at the same time who all survived is eight. They were born to Nadya Suleman on 26 January 2009, after IVF treatment in which 12 embryos were placed in her uterus at the same time.

1 For each of the following hormones, identify where it is released and its target organ(s).

6th **a** oestrogen

6th **b** progesterone

6th **c** FSH

6th **d** LH

2 Describe how the four hormones interact to bring about:

8th **a** ovulation

8th **b** menstruation.

FSH (follicle-stimulating hormone) and **LH** (luteinising hormone) are released from the pituitary gland. The release of these hormones is controlled by the concentration of oestrogen (which increases as the **egg follicle** matures) and of progesterone (which is released after ovulation when the follicle becomes a structure called the **corpus luteum**). Diagram B shows how the four hormones interact to control the menstrual cycle.

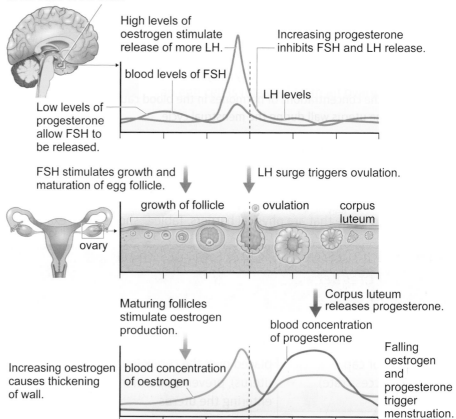

B Many hormones interact to control the menstrual cycle.

Hormonal contraception uses a progesterone-like hormone either on its own or with oestrogen. Raising hormone concentrations in this way prevents the natural fall of concentrations at the end of the menstrual cycle.

8th **3** Explain how the effect of hormonal contraception on FSH and LH can help to prevent pregnancy.

H

There are many possible reasons why some couples are unable to have a child. Some problems can be overcome using **Assisted Reproductive Technology (ART)**, which uses hormones and other techniques to increase the chance of pregnancy.

Clomifene therapy is useful for women who rarely or never release an egg cell during their menstrual cycles. Clomifene is a drug that helps to increase the concentration of FSH and LH in the blood.

 4 Explain how clomifene works to increase the chance of pregnancy for some women.

Another ART technique is **IVF** (in vitro fertilisation). This can overcome problems such as blocked oviducts in the woman, or if the man produces very few healthy sperm cells. Diagram C shows how IVF is carried out. Any healthy embryos not used in the first attempt at pregnancy may be frozen and stored for use another time.

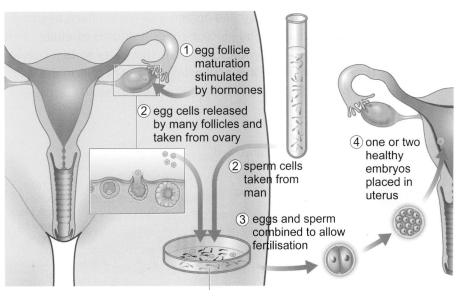

① egg follicle maturation stimulated by hormones

② egg cells released by many follicles and taken from ovary

② sperm cells taken from man

③ eggs and sperm combined to allow fertilisation

④ one or two healthy embryos placed in uterus

The technique is called 'in vitro' (which means 'in glass') because glass dishes were used originally.

C how IVF is carried out

 5 a Explain why the woman's ovaries are stimulated with hormones at the start of IVF treatment.

 b Suggest which of the hormones that normally control the menstrual cycle is given to stimulate the release of egg cells in IVF treatment. Explain your reasoning.

Exam-style question

Explain how a hormonal treatment using progesterone can prevent pregnancy.
(3 marks)

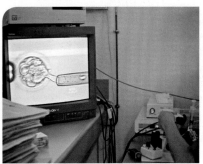
Checkpoint

How confidently can you answer the Progression questions?

Strengthen

S1 If an egg cell is fertilised, the corpus luteum remains large and active. Explain why a missed period is often the first sign of pregnancy.

Extend

E1 Clomifene therapy works by blocking the negative feedback effect of oestrogen on a pituitary hormone involved in the control of the menstrual cycle. Explain what this means.

SB7e Control of blood glucose

Specification reference: B7.9; B7.13; H B7.14; B7.15

Progression questions

- What is homeostasis?
- How is blood glucose concentration regulated?
- How can type 1 diabetes be controlled?

A A 15th century urine colour chart helped doctors to diagnose illnesses in their patients.

Urine tests are a simple way of testing for pregnancy or for many diseases including **diabetes**. The substances in urine can provide important clues. However, while a doctor in the Middle Ages might have tasted the urine to test for diabetes, doctors today can use simple chemical tests.

During digestion in the gut, glucose is released from carbohydrates in our food. Glucose is easily absorbed from the small intestine into the blood and then into cells, where it is broken down during respiration.

It takes time for cells to take in the glucose released by digestion, so there is a risk that glucose may reach a very high concentration in the blood. This is dangerous because it can damage organs. However, in most people this does not happen, because blood glucose concentration is carefully controlled. As blood glucose concentration rises, it stimulates certain cells in the pancreas to release the hormone **insulin**. Insulin causes cells in the liver and other organs to take in glucose, which causes a fall in blood glucose concentration.

6th **1** **a** What happens to glucose during respiration?

6th **b** Explain why glucose is important for the body.

6th **2** **a** What will happen to the concentration of glucose in the blood soon after a meal?

7th **b** Explain your answer to part a.

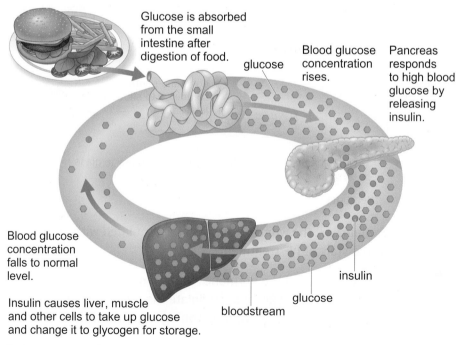

Glucose is absorbed from the small intestine after digestion of food.

Blood glucose concentration rises.

Pancreas responds to high blood glucose by releasing insulin.

glucose

Blood glucose concentration falls to normal level.

Insulin causes liver, muscle and other cells to take up glucose and change it to glycogen for storage.

insulin

glucose

bloodstream

B the control of glucose concentration in the blood

6th **3** Describe how the body responds when blood glucose concentration rises above normal.

As blood glucose concentration falls, the insulin-releasing cells in the pancreas release less and less insulin. If blood glucose concentration falls below a certain level, the cells stop releasing insulin altogether.

H

When glucose is absorbed by the liver, it is converted to glycogen, which is stored in the liver cells. If blood glucose concentration falls too low, another hormone is released from other pancreatic cells. This hormone is called **glucagon**. Glucagon causes liver cells to convert glycogen back to glucose, which is released into the blood. As blood glucose concentration increases, the amount of glucagon released from the pancreas falls.

4 Compare the roles of glucagon and insulin in the control of blood glucose concentration.

5 Explain why the control of blood glucose concentration is an example of negative feedback. It may help to draw a diagram.

The effects of hormones help to keep blood glucose concentration within limits that are safe. Maintaining constant conditions inside the body is called **homeostasis**. Other examples of homeostasis in the body include temperature control and the control of water content. All the processes involved in homeostasis help to prevent damage to the body as internal and external conditions change.

6 Explain how the homeostatic control of blood glucose concentration helps to protect the body.

Type 1 diabetes

In a few people, the pancreatic cells that should produce insulin do not. This is because the cells have been destroyed by the body's immune system. These people have **type 1 diabetes**, and it means that they cannot control rising blood glucose concentration.

When blood glucose concentration is too high, some glucose can be detected in the urine. Therefore, glucose in the urine is often the first test for type 1 diabetes.

People with type 1 diabetes have to inject insulin into the fat layer below the skin, where it can enter the blood, causing blood glucose concentration to fall.

7 Explain how the following may affect how much insulin someone with type 1 diabetes should inject.

a time since last meal

b the types of food eaten in a meal

c the amount of recent exercise

Exam-style question

Explain why people with type 1 diabetes must control their blood glucose concentration with injections of insulin. *(2 marks)*

C In order to decide on the right amount of insulin to inject, a person with type 1 diabetes often does a simple blood test to check their blood glucose concentration.

Checkpoint

How confidently can you answer the Progression questions?

Strengthen

S1 Using an example, explain why homeostasis is important.

Extend

E1 **H** The control of blood glucose concentration involves two hormones. Suggest how this can provide better control of concentration than using just one hormone.

SB7g Thermoregulation

Specification reference: B7.10B; B7.11B; B7.12Ba; **H** B7.12Bb, c

Progression questions

- Why is it important to control core body temperature?
- How are the skin, muscles and the hypothalamus involved in controlling body temperature?
- **H** How do blood vessels help in controlling body temperature?

A The casualty in this rescue has been wrapped in insulating layers to help raise his body temperature. Otherwise he could become unconscious or even die from hypothermia.

The normal temperature of the major organs (heart, liver and brain) of the human body is about 37 °C, although this varies slightly between people. A temperature above 38 °C is a **fever**, and below 36 °C causes **hypothermia**. Both fever and hypothermia are dangerous because they affect how well the enzymes in the body work. **Thermoregulation** is the control of body temperature, which keeps the temperature of the major organs close to 37 °C most of the time.

 1 Control of body temperature is another example of homeostasis. State what this means.

 2 **a** Describe the effect of temperature on the rate of an enzyme-controlled reaction.

 b Explain why temperature affects how well enzymes in the major organs work.

The **hypothalamus** is a small part of the brain that constantly monitors temperature. It receives information from temperature receptors in the **dermis** of the skin (see diagram D). Receptors inside the hypothalamus detect temperature changes in the brain and the blood.

If the hypothalamus detects blood or brain temperatures starting to fall below 37 °C or it detects a cold environment (that could cause body temperature to fall), it causes various changes.

- **Shivering** is when muscles start to contract and relax rapidly. Some of the energy released from cell respiration for shivering warms you up.
- Contraction of **erector muscles** in the dermis of the skin causes body hairs to stand upright. In humans this has little effect, but in other mammals it traps air next to the skin for insulation.
- Reduction of blood flow near the skin keeps warm blood deeper inside the body. This reduces the rate of transfer of energy to the air by heating.

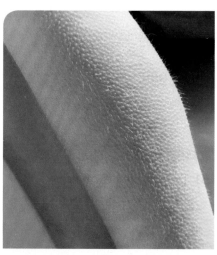

B Goosebumps are a sign that the body is too cold. The bumps are caused by the contraction of hair erector muscles that raise the hairs.

 3 **a** Explain how the hypothalamus detects a cold environment.

 b Explain how the hypothalamus detects a body temperature that is too low.

 4 Explain why shivering helps you to warm up.

If body temperature rises above 37 °C, the hypothalamus detects this and causes sweating. Sweat spreads out as a thin layer over the skin **epidermis**, where it evaporates. As sweat evaporates it transfers energy from the skin to the surroundings by heating, so the skin cools down. The hypothalamus also increases blood flow nearer to the surface of the skin. This makes it easier for the blood to transfer energy to the air, so we cool down.

 5 Explain two changes in the skin that help to cool the body down when it is too warm.

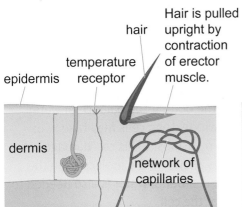

D how the skin responds when the core body temperature is too low (left), and when core body temperature is too high (right)

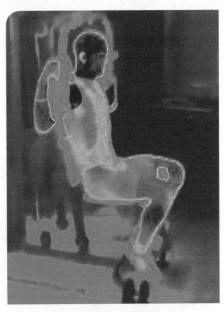

C This thermal image shows the increase of blood flow bringing hot blood closer to the surface of the skin during exercise.

H

When it is cold, the hypothalamus sends nerve impulses to small arteries deep in the skin, causing them to narrow. This narrowing of blood vessels is called **vasoconstriction**. This reduces blood flow in capillaries near the surface of the skin and helps to reduce energy transfer to the surroundings. When the body is hot, the hypothalamus causes the small arteries to widen (**vasodilation**). This increases blood flow through skin capillaries, bringing warm blood nearer to the surface of the skin and increasing energy transfer to the surroundings.

The control of body temperature is another example of **negative feedback**. This helps keep conditions in the body under control at around the right level.

 6 Use diagram D to help you explain how changes in skin blood vessels help to cool the body when it is too hot.

 7 Explain why thermoregulation is an example of control by negative feedback.

Checkpoint

How confidently can you answer the Progression questions?

Strengthen

S1 Draw a sketch to show the changes in the body when body temperature falls too low.

Extend

E1 Explain the importance of covering the body of the man in photo A.

Exam-style question

Describe the monitoring and control roles of the hypothalamus in thermoregulation. *(3 marks)*

SB7h Osmoregulation

Specification reference: B7.10B; B7.18B; B7.21B; B7.22B

Progression questions

- Why is osmoregulation important?
- What is the structure of the urinary system?
- How can kidney failure be treated?

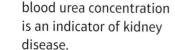

A Food and drink stations are usually placed at regular intervals in ultramarathons, to help runners reduce the risk of harm to their kidneys and other organs.

An ultramarathon is a running race over a distance of 50 km or more. These are tests of endurance, and competitors need to manage their fluid intake carefully. Too much or too little water drunk before and during the race can lead to kidney damage.

Osmoregulation is the control of the balance of water and mineral salts in the body. If the balance of water and mineral salts is wrong, then cells may take in or lose too much water by osmosis. This can damage cells because water in cells allows all the molecules in the cell's reactions to move around. Water is also needed to maintain the shape of the cell.

 1 Explain why osmoregulation is important.

The function of the **urinary system** is to remove excess amounts of some substances from the blood, including water and mineral salts. It also removes waste products, such as **urea**. Urea is produced in liver cells from the breakdown of amino acids that are in greater amounts than are needed. The urea passes into the blood and is carried to the **kidneys**.

The renal veins carry blood with wastes removed back to the body.

The ureters carry urine from the kidneys to the bladder.

The bladder stores urine.

Urine flows through the urethra to the outside of the body.

The renal arteries carry blood from the body to the kidneys.

The kidneys remove substances from the blood and make urine.

A muscle keeps the exit from the bladder closed until a person decides to urinate.

B structure of the human urinary system

 2 Explain why an increase in blood urea concentration is an indicator of kidney disease.

3 Use diagram B to:

a identify the blood vessels that carry urea to the kidneys

b describe how substances removed from the blood pass to the outside of the body.

Your body can manage with only one healthy kidney. Sometimes (e.g. due to an infection) a person will suffer **kidney failure**, when both kidneys stop working properly. The person's life will be in danger, because waste substances increase in concentration in the blood. The person will need kidney **dialysis** every few days to keep the concentration of substances in the blood at safe levels. Diagram C shows how a dialysis machine works.

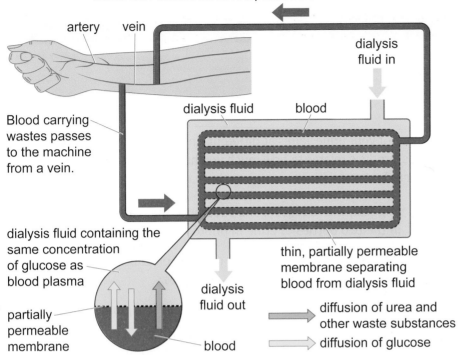

Blood with wastes removed passes from the machine to a vein.

artery   vein

dialysis fluid in

Blood carrying wastes passes to the machine from a vein.

dialysis fluid   blood

dialysis fluid containing the same concentration of glucose as blood plasma

thin, partially permeable membrane separating blood from dialysis fluid

dialysis fluid out

partially permeable membrane

blood

→ diffusion of urea and other waste substances

⇨ diffusion of glucose

C During dialysis, substances are exchanged between the blood and dialysis fluid by diffusion across the dialysis membrane (which is partially permeable to very small molecules such as water, glucose and urea).

A better treatment for kidney failure is **organ donation**, when a kidney from another person is put into a patient's body and attached to their blood system. Replacing a kidney involves several hours of surgery, which may be too much for weak patients. So donation is not suitable for all. Also, kidney cells, like all cells, have **antigens** on them. Cells in the immune system recognise and attack strange antigens. This can cause **rejection** of the donated kidney. The antigens on the donated organ must therefore be matched to those on the patient's cells. It can take a long time to find a suitable kidney for a patient.

Even with a good match, the patient will need life-long medication to prevent the kidney being rejected. This medication affects the body's response to infection, so the patient may catch infections more easily.

5 Explain why someone who receives dialysis must be treated every few days.

6 Explain why not all patients with kidney failure are given a donated kidney.

7 Suggest why organ donation can give better treatment for kidney failure than dialysis.

 4 Look at diagram C. State, with a reason, whether the concentration of urea in the dialysis fluid should be higher or lower than in the blood at the start of treatment.

Did you know?

Most donated kidneys are from people who have just died, such as in car accidents, but who had healthy kidneys. A patient match must be found quickly as the kidneys remain healthy for only around 30 hours after death when kept properly.

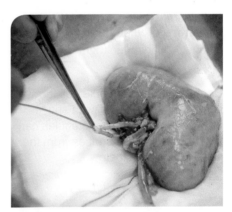

D This kidney has been taken from a donor and is being prepared for transplant.

Checkpoint

How confidently can you answer the Progression questions?

Strengthen

S1 Explain why someone who has kidney failure needs to be treated.

Extend

E1 Suggest, with a reason for your suggestion, how you could easily monitor the health of a donated kidney.

Progression questions

- What are the parts of a nephron?
- How does filtration and reabsorption take place in a nephron?
- H How does ADH affect nephrons?

A After drinking a lot, you usually produce large quantities of pale-coloured urine (left bottle). If you are **dehydrated**, you produce only a small amount of dark urine (right bottle).

Each Bowman's capsule cell forms a vast series of 'fingers' with spaces between them. Small molecules can filter through these spaces from the 'leaky' capillaries below.

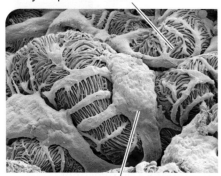

main part of Bowman's capsule cell

B The capillary cells in the glomerulus and the cells that form the Bowman's capsule have spaces between them. This makes both the glomerulus and the Bowman's capsule leaky.

 2 **a** Describe what happens in kidney filtration.

 b Explain how the kidney is adapted for filtration.

The colour of your urine varies at different times of day, and the quantity of urine you produce relates to how much you have drunk or how active you have been. These variations are due to the way your kidneys work.

 1 Suggest, with a reason, one substance that varies in concentration in the two bottles of urine in photo A.

Each kidney contains thousands of tiny microscopic tubes called **nephrons**. **Urine** is made in the nephron in a series of stages.

- Blood flows through a network of capillaries called a **glomerulus**, which runs inside the **Bowman's capsule** of each nephron.
- The Bowman's capsule and glomerulus are adapted to let very small molecules, such as water, urea and glucose, through into the nephron. Large molecules such as proteins, and blood cells, stay in the blood. This process is called **filtration**.

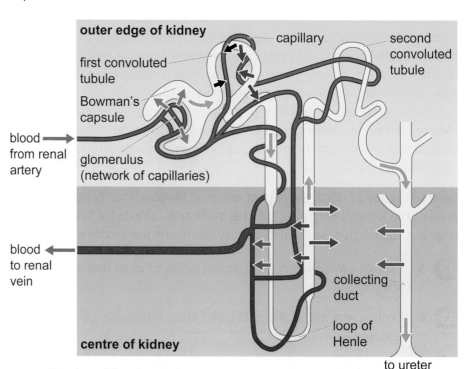

C the structure of a nephron

- The filtration fluid flows along inside the nephron. **Selective reabsorption** of useful substances that the body needs occurs here. This includes glucose and some mineral ions. These substances are pumped through proteins in the cell membranes in the **first convoluted tubule** of the nephron, by **active transport**. No glucose is normally left in urine.
- Water is reabsorbed by osmosis, depending on how much the body needs. This happens in the **loop of Henle** and in the **collecting duct**.
- At the end of the nephron the remaining fluid flows into the ureter. The fluid contains excess water that the body does not need, plus urea and other substances. It is now called urine.

The nephron is adapted in several ways for reabsorption of substances.

- There is a large surface area of contact between the nephron and capillaries.
- The cell membrane of the cells lining the first convoluted tubule has tiny folds called **microvilli**. These increase the surface area : volume ratio of the cells.
- Cells that have protein pumps in their cell membranes contain many mitochondria.

 3 Describe what is meant by selective reabsorption, using an example in your answer.

 4 Explain why cells involved in the selective reabsorption of glucose contain many mitochondria.

 5 Explain how the close relationship between a nephron and a capillary supports reabsorption.

H Controlling water content

As well as getting rid of urea, the kidneys help to control the water content of the blood. When the **pituitary gland** detects that there is too little water in the blood, it releases the hormone **ADH** (antidiuretic hormone). ADH changes the **permeability** of the collecting duct in nephrons and increases the concentration of the urine, as shown in diagram D.

If the collecting duct is permeable, water is absorbed by osmosis from the collecting duct back into the blood. When there is plenty of water in the blood, the pituitary gland stops releasing ADH.

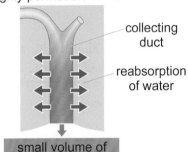

D ADH changes the permeability of the collecting duct and so the amount of urine formed.

 6 Explain how ADH affects urine formation.

 7 Explain how the control of the body's water content is an example of negative feedback.

Did you know?

Porphyria is a genetic disorder that causes purple urine. A faulty enzyme results in a build-up of a red-coloured substance needed to make haemoglobin molecules. The excess of this substance is excreted in the urine in concentrated form.

Exam-style question

Describe how filtration and reabsorption in the kidneys lead to the formation of urine.

(2 marks)

Checkpoint

How confidently can you answer the Progression questions?

Strengthen

S1 Explain why urine normally contains urea but not glucose.

Extend

E1 H Suggest the difference in concentration of ADH in blood that produced each of the two urine samples in photo A. Explain your answer.

Risk of type 2 diabetes

A Swedish study investigated waist:hip ratios and type 2 diabetes. The waist:hip ratios of 792 54-year-old men in Gothenburg, Sweden were recorded. Thirteen years later, the scientists recorded whether or not these men had developed type 2 diabetes.

The waist:hip ratio values were arranged in order of size and then split into three groups: small, medium and large waist:hip ratios.

The mean probability of developing type 2 diabetes was calculated for each group. The chart shows the results.

Evaluate the relationship between waist:hip ratio and the risk of developing type 2 diabetes, using the results shown in the chart.

(6 marks)

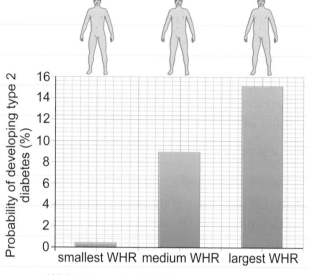

Waist:hip ratio (WHR) groups at start of study

Student answer

Between the smallest and the medium WHRs the probability of developing type 2 diabetes goes up to 8.6% and then goes up another 6.2% between the medium WHR and the largest WHR. So it looks like the higher someone's WHR, the more likely they are to develop type 2 diabetes [1].

The study included 792 men, which is a large number. This reduces the possibility that the results have happened by chance, and so suggests they could be repeated in similar studies [2].

The study only included men from one Swedish city who were 54 years old at the start of the study. Different groups of people, of different ages, gender or place might show different results [3].

This means that you cannot judge if the trend shown in the chart is true for everyone or just for the men who took part in the study [4].

[1] It is essential that you quote and use actual data from a graph or chart in questions of this type. This student has done this here, and used the data to then state the overall pattern shown.

[2] This part of the answer clearly identifies a strength of the study, and explains why it is a strength.

[3] This part of the answer clearly identifies a weakness of the study, and explains why it is a weakness.

[4] The answer finishes well with a conclusion based on the strength and weakness identified.

Verdict

This is a strong answer. It clearly explains what the chart shows. Some strengths and weaknesses of the study are set out in a logical and well-structured way, making it easy to follow the discussion, and to understand the conclusion.

Exam tip

'Evaluate' questions require you to consider the strengths and weaknesses of information and form a conclusion that takes those into account. It is a good idea to plan these questions by listing 'strengths' and 'weaknesses' before you start writing your answer.

Paper 2

SB8 Exchange and Transport in Animals

This shark has had all its skin and flesh removed. What you can see is just the dense network of capillaries, arteries and veins that forms its circulatory system. There are so many blood vessels because the system needs to ensure that every single cell in the shark's body gets an adequate supply of oxygen and food. The same is true for all large animals, including humans. In this unit you will find out about the importance of the circulatory system and how it works.

The learning journey

Previously you will have learnt at KS3:

- how the digestive system gets glucose and other food molecules into the blood
- how the respiratory (breathing) system gets oxygen into the blood
- about aerobic and anaerobic respiration.

You will also have learnt in *SB1 Key Concepts in Biology*:

- about diffusion
- about different animal cells and their adaptations.

In this unit you will learn:

- more about diffusion, gas exchange and the surface area : volume ratio
- about the rate of diffusion and Fick's law
- more about the different types of respiration
- how the lungs, heart, blood vessels and blood are adapted for their functions
- how to calculate cardiac output.

SB8a Efficient transport and exchange

Specification reference: B8.1; B8.2; B8.3

Progression questions

- What substances need to be transported into and out of the body?
- Why is the surface area:volume ratio important for exchange of substances?
- How are lungs adapted for gas exchange?

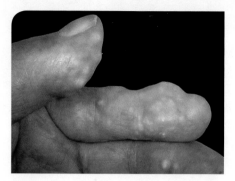

A Waste DNA is broken down into uric acid and excreted by your kidneys. If it builds up in the blood it can cause gout, in which uric acid crystals form and cause painful swellings.

All the chemical reactions in your body (your **metabolism**) produce wastes, which must be **excreted** so they do not cause problems. Your kidneys remove **urea**, which is a poison produced by breaking down amino acids. Your lungs get rid of carbon dioxide produced in **aerobic respiration**.

 1 Name two human excretory organs.

Your body also moves substances into it. Oxygen and glucose are needed for aerobic respiration. Dissolved food molecules (e.g. glucose, amino acids) and mineral ions are needed to produce new substances for your body.

 2 a What large molecules are made using amino acids?

 b When these molecules are broken down again, what waste do they form?

 c How is this waste excreted from the body?

 3 What organs take the substances needed for aerobic respiration into the body?

Small blood vessels, called capillaries, have walls just one cell thick.

Oxygen molecules (not shown) diffuse out of the capillary.

Glucose molecules (not shown) diffuse out of the capillary.

The continual flow of blood maintains the concentration gradient.

nucleus

There is an overall movement of carbon dioxide molecules down their concentration gradient.

B Substances diffuse down their concentration gradients into and out of narrow blood vessels called capillaries.

Many substances move into and out of parts of the body by **diffusion**. To make sure a lot of particles diffuse quickly, the surfaces through which they move:

- are thin – so that particles do not need to diffuse very far
- have a large surface area – so that there is more room for particles to diffuse.

 4 a In diagram B, is the glucose concentration highest inside or outside the vessel? Explain your reasoning.

 b Why do oxygen molecules diffuse out of the capillary?

 5 Describe one way in which a capillary is adapted so that substances diffuse quickly in and out.

Did you know?

Capillaries are 5–10 μm in diameter, with walls that are about 0.6 μm thick.

Surface area:volume ratio

It would take too long for materials to diffuse through cells on the outside of a tissue to reach cells on the inside. So, **multicellular organisms** have transport systems. In humans, a fine network of **capillaries** in the **circulatory system** uses blood to transport substances to and from all cells.

The larger a cell's surface area, the more of a substance can diffuse into (and out of) it in a certain time. However, if a cell's volume is too big, the cell cannot fill up with all the materials it needs quickly enough.

The **surface area : volume ratio** (**SA : V**) is the surface area divided by the volume, or

$$\frac{\text{surface area}}{\text{volume}}$$

The bigger this ratio, the more surface area something has per unit volume. Diagram C shows that as cells get bigger, their SA : V ratio gets smaller. If the ratio is too small, a cell cannot get enough raw materials fast enough. So, there is a limit to the size of cells.

Organs that move substances into and out of the body have large SA : V ratios. For example, a human lung has about the same volume as a football, but its surface area is about 250 times greater. This is because lungs are packed with millions of **alveoli**, which increase the surface area and so increase the speed and amount of **gas exchange**.

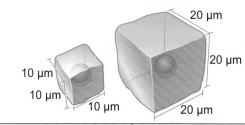

surface area	= 6 × (10 × 10) = 600 µm²	surface area	= 6 × (20 × 20) = 2400 µm²
volume	= 10 × 10 × 10 = 1000 µm³	volume	= 20 × 20 × 20 = 8000 µm³
SA : V	= $\frac{600}{1000}$ = 0.6	SA : V	= $\frac{2400}{8000}$ = 0.3

C Cells of different sizes have different SA : V ratios.

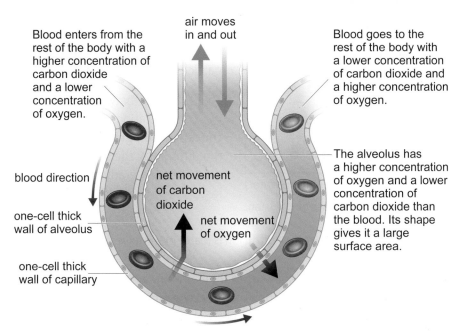

Blood enters from the rest of the body with a higher concentration of carbon dioxide and a lower concentration of oxygen.

air moves in and out

Blood goes to the rest of the body with a lower concentration of carbon dioxide and a higher concentration of oxygen.

blood direction

net movement of carbon dioxide

net movement of oxygen

The alveolus has a higher concentration of oxygen and a lower concentration of carbon dioxide than the blood. Its shape gives it a large surface area.

one-cell thick wall of alveolus

one-cell thick wall of capillary

D The drawing shows an alveolus, which is adapted for fast gas exchange (swapping of gases). An adult lung contains about 500 million alveoli, which are grouped together in clusters at the ends of tiny tubes.

 8 Explain how alveoli are adapted for fast gas exchange.

6 A skin cell is a cube with sides of 3 µm. Calculate:

 a its volume

 b its surface area

 c its SA : V ratio.

7 Why is there a limit to cell size in a multicellular organism?

Exam-style question

Explain why humans need a circulatory system. *(2 marks)*

SB8b Factors affecting diffusion

Specification reference: B8.4B; B8.5B

Progression questions

- How do surface area and concentration affect the rate of diffusion?
- What is the relationship between the rate of diffusion and diffusion distance?
- What is Fick's law?

1 20 g of glucose is dissolved in 160 cm³ of water. Calculate the glucose concentration in:

a $g\ cm^{-3}$

b $g\ dm^{-3}$.

A **concentration** is the amount of a substance in a certain volume. A common unit is g/cm³ or g cm⁻³, where the small minus sign shows that g is divided by cm³. Another common unit is g dm⁻³ (1 dm = 1 litre = 1000 cm³).

You can calculate the concentration of a solution in g dm⁻³ using this equation:

$$\text{concentration} = \frac{\text{mass of solute in g}}{\text{volume of solution in dm}^3}$$

The number of particles *decreases* as you go *down* a concentration gradient.

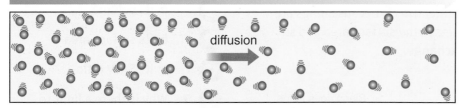

diffusion

The steeper the concentration gradient, the faster the rate of diffusion.

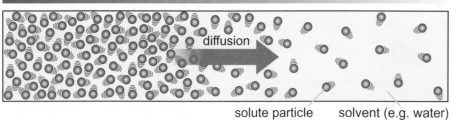

diffusion

solute particle   solvent (e.g. water)

A The greater the difference between concentrations, the faster the rate of diffusion.

The particles in a solution move randomly in all directions. This causes an overall ('net') movement of the solute particles, from higher concentration to lower concentration. No net movement occurs when the concentrations are equal (although the individual particles are still moving).

The difference between two concentrations forms a **concentration gradient**. The bigger the difference, the *steeper* the concentration gradient and the faster the rate of diffusion.

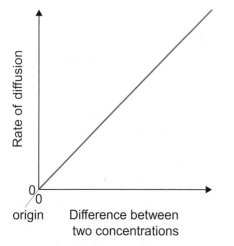

B The rate of diffusion is directly proportional to the concentration difference.

 2 Explain why the rate of diffusion is faster in the lower part of diagram A.

The straight line on graph B shows that there is a **linear relationship**. The line goes through the origin, which shows a **directly proportional** relationship (as one variable increases by a certain percentage, the other increases by the same percentage). We write this by using the 'proportional to', ∝, symbol:

rate of diffusion ∝ concentration difference

To keep the rate of diffusion high, a steep concentration gradient must be maintained. In the lungs, a good blood supply moves oxygen quickly out of the lungs. This maintains the concentration gradient.

 3 The oxygen concentration at place X = 1000 molecules/μm³, at Y = 5000 molecules/μm³ and at Z = 4000 molecules/μm³. Between which two is the rate of diffusion highest? Explain your answer.

4 Look back at diagram B in SB8a.

 a How is the concentration gradient in the capillary maintained?

 b Why is this useful?

Surface area

Small particles can pass through membranes in organisms. If the surface area of a membrane is increased, there is more space through which particles can pass. This means that more particles cross from one place to another in a certain time, and so the overall rate of diffusion increases (but the rate at which particles pass through each unit area of the surface membrane is unchanged).

rate of diffusion ∝ surface area

 5 Describe the shape of the line on a graph which shows the rate of diffusion against surface area.

6 Oxygen diffuses through $10\,cm^2$ of an alveolus wall at $0.01\,\mu g/s$.

 a Calculate the overall rate of diffusion if the surface area is tripled.

 b Calculate the overall rate of diffusion if the surface area is halved.

 c Calculate the rate of diffusion per cm^2 for the conditions in **a** and **b**.

Distance

The further particles have to diffuse, the slower the rate of diffusion. So *increasing* the thickness of a membrane *decreases* the rate of diffusion. This is an **inversely proportional** relationship. As one variable doubles, the other halves.

$$\text{rate of diffusion} \propto \frac{1}{\text{thickness of membrane}}$$

 7 Carbon dioxide diffuses into a capillary at 1500 molecules/s. Calculate the rate of diffusion if the capillary wall thickness doubles.

Fick's law shows the relationship between the variables that affect diffusion:

$$\text{rate of diffusion} \propto \frac{\text{surface area} \times \text{concentration difference}}{\text{thickness of membrane}}$$

 8 Carbon dioxide diffuses into an alveolus at $0.0001\,g/s$. Calculate the rate of diffusion if the:

a surface area is reduced by 30 per cent

b concentration difference is increased by 20 per cent

c alveolus wall thickness is halved.

Exam-style question

Describe what Fick's law shows. *(3 marks)*

We can imagine a membrane as having gaps in it through which particles can pass.

slower diffusion　　　faster diffusion

C The larger the surface area, the faster the rate of diffusion.

Did you know?

All gases need to be dissolved in water in order to diffuse into and out of the body (which is why the surfaces of the lungs are wet).

Checkpoint

How confidently can you answer the Progression questions?

Strengthen

S1 Draw a table to show how different variables affect diffusion.

Extend

E1 Glucose diffuses into a part of the small intestine at $0.001\,mg/s$. Due to a disease, the small intestine becomes twice as thick and loses one-third of its surface area. Use Fick's law to calculate the effect of these changes on the rate of glucose diffusion.

SB8c The circulatory system

Specification reference: B8.6; B8.7

Progression questions

- What are the components of the circulatory system?
- How are blood vessels adapted to their functions?
- How is blood adapted to its function?

network of fine capillaries in the lungs

veins carry blood back to the heart

wide tube

thin, flexible wall

heart

arteries take blood away from the heart

narrow tube

thick layer of elastic and musle fibres

capillaries in tissues
wall is only one cell thick, to allow faster diffusion of substances into and out of the capillary

very narrow tube

A the human circulatory system

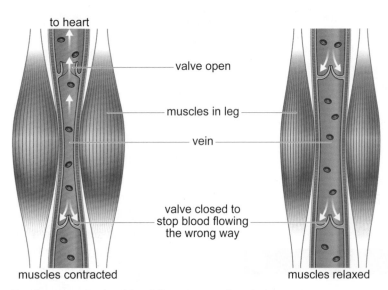

to heart

valve open

muscles in leg

vein

valve closed to stop blood flowing the wrong way

muscles contracted

muscles relaxed

B Valves ensure that blood flows in one direction.

In the circulatory system, **blood** flows away from the **heart** into **arteries**. These divide into narrow **capillaries**, which form fine networks running through tissues. Blood returns to the heart in **veins**.

 1 List the parts of the circulatory system.

 2 a Describe how the surface area for the exchange of substances is made so large in tissues.

 b Explain why this is necessary.

With each beat, the heart squirts blood into arteries under high pressure. Artery walls are thick to withstand this sudden increase in pressure, but it makes them stretch. A wave of stretching then passes along the artery walls. You feel this wave as a **pulse** (the pulse is not your blood moving).

After stretching, muscle and elastic fibres in the artery walls cause the arteries to contract again. The stretching and contracting of arteries makes the blood flow more smoothly.

Blood flows under low pressure in veins and so they only need thin walls. As you move, muscles in your skeleton help to push blood along the veins. Veins contain **valves** to prevent blood flowing the wrong way (as shown in diagram B).

 3 Draw a table to show one way in which each type of blood vessel is adapted for its function.

 4 Explain why your pulse rate is the same as the rate at which your heart beats.

 5 Some people have a disease in which the vein valves in their legs do not work properly. Suggest one symptom of this disease.

Blood

In each cubic millimetre (mm³) of blood there are about 5 000 000 **erythrocytes** (**red blood cells**), 7000 **white blood cells** and 250 000 **platelets**. The cells are suspended in a straw-coloured liquid called **plasma**, which carries dissolved substances such as glucose, carbon dioxide and urea.

Red blood cells are packed with **haemoglobin**. This substance binds with oxygen in the lungs and releases it again in tissues. When a lot of oxygen is bound to haemoglobin molecules, the cells are bright red. When there is less oxygen attached to the molecules, the cells are dark red.

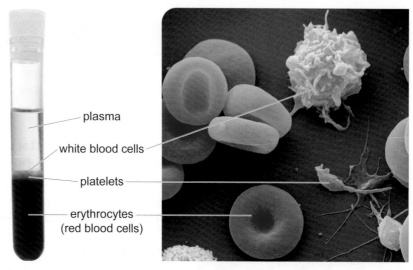

plasma

white blood cells

platelets

erythrocytes (red blood cells)

C the components of blood

Erythrocytes have no nucleus, so there is more space for haemoglobin. The cells are shaped like discs with a dimple in each side. This 'biconcave' shape allows a large surface area : volume ratio for oxygen to diffuse in and out.

There are different types of white blood cells, including **phagocytes** and **lymphocytes**, which remove foreign cells that get inside you. Lymphocytes produce proteins called **antibodies** that stick to foreign cells and help to destroy them. Phagocytes surround foreign cells and digest them.

Platelets are tiny fragments of cells that have no nuclei. Platelets produce substances needed to clot the blood at the site of an injury, for example when the skin is cut.

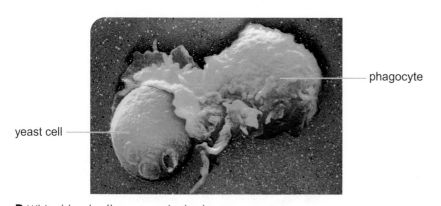

phagocyte

yeast cell

D White blood cells protect the body.

8 Write a better caption for photo D to describe what the photo shows.

9 Look at diagram D on *CB8a Efficient transport and exchange*. Explain why the erythrocytes are not all the same colour.

6 State the difference between the way urea and oxygen are carried in the blood.

7 Explain how erythrocytes are adapted to their function.

Checkpoint

How confidently can you answer the Progression questions?

Strengthen

S1 Draw a table to contrast arteries and veins. Include their structures and the substances in the blood inside them.

Extend

E1 In a condition called atherosclerosis, fatty substances build up inside arteries. This makes the arteries narrower and their walls harder. Explain the effects that this condition might have on someone.

Exam-style question

Explain how oxygen is transported from the lungs to a tissue.   *(3 marks)*

SB8d The heart

Specification reference: B8.8; B8.12

Progression questions

- What is the structure of the heart like?
- How does the heart pump blood?
- How do you calculate cardiac output?

A This prototype drone can quickly reach a heart attack patient and deliver a defibrillator.

 1 Suggest why heart muscle dies if it does not get blood.

Someone has a **heart attack** in the UK every three minutes. A heart attack occurs when blood stops flowing to muscles in part of the heart, damaging them and stopping the heart pumping properly. If the heart stops completely, it can often be started again by putting an electric shock through it (using a defibrillator).

Heart structure

There are four **chambers** in the heart. Blood from most of the body enters the right **atrium** through the **vena cava** (a large vein). At the same time, blood from the lungs enters the left atrium through the **pulmonary vein**. When these top chambers are full, the muscles around them **contract** to push blood into the **ventricles**. The muscles in the ventricle walls then contract, forcing blood out of the heart. As this is happening, the muscles in the atria walls relax and these chambers refill with blood.

Heart valves stop blood flowing the wrong way. It is the sound of these valves shutting that you hear as 'lub-dub' when listening to a heart.

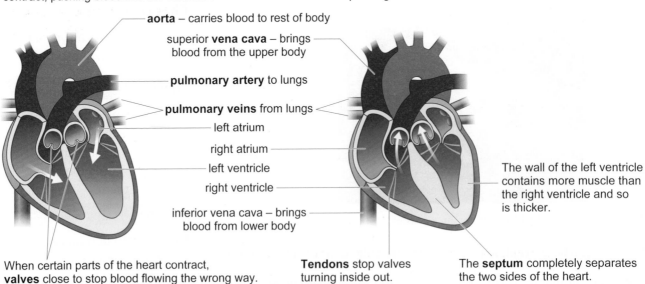

Blood flows into the two **atria**, which then contract, pushing blood into the ventricles.

The **ventricles** then contract, pushing blood out of the heart.

aorta – carries blood to rest of body

superior **vena cava** – brings blood from the upper body

pulmonary artery to lungs

pulmonary veins from lungs

left atrium

right atrium

left ventricle

right ventricle

inferior vena cava – brings blood from lower body

The wall of the left ventricle contains more muscle than the right ventricle and so is thicker.

When certain parts of the heart contract, **valves** close to stop blood flowing the wrong way.

Tendons stop valves turning inside out.

The **septum** completely separates the two sides of the heart.

B A heart is always drawn as though the person were facing you.

Heart muscle is all the same colour, but in diagram B the parts containing blood with little oxygen (**deoxygenated** blood) are coloured dark red. The parts that pump **oxygenated** blood are coloured bright red.

2 a How many atria does the heart contain?

 b Why is the left ventricle on the right of the heart in diagram B?

 3 List in order the parts through which blood flows from the vena cava to the aorta.

 4 Explain why blood in the vena cava is dark red but blood in the aorta is bright red.

 5 The left ventricle has a thicker muscle wall, and so contracts more strongly than the right. Explain why this is needed.

 6 What do the heart valves do?

Cardiac output

The contraction and relaxation of muscles during each heartbeat is controlled by **impulses** from the nervous system. The **heart rate** is the number of times the heart beats in a minute. The volume of blood pushed into the aorta in each beat is the **stroke volume**. It is measured in litres. The **cardiac output** is the volume of blood pushed into the aorta each minute, and can be calculated using the equation:

cardiac output = stroke volume × heart rate
(litres/min)   (litres/beat)   (beats/min)

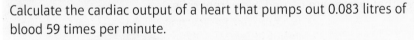

Worked example

Calculate the cardiac output of a heart that pumps out 0.083 litres of blood 59 times per minute.

cardiac output = stroke volume × heart rate

cardiac output = 0.083 × 59 = 4.9 litres/min (to 2 significant figures)

Regular exercise increases the strength of heart muscle and ventricle size. So, fitter people often have bigger stroke volumes, and their hearts can beat more slowly to achieve the same cardiac output as a less fit person.

 7 a Calculate the cardiac output of a heart that pumps 0.07 litres of blood 55 times per minute.

 b Calculate the stroke volume for a cardiac output of 5 litres/min and a heart rate of 50 beats/min.

 8 Explain why people who take regular exercise often have slower heart rates than those who do not.

Exam-style question

Suggest an explanation for why the heart is sometimes called a 'double pump'. *(2 marks)*

Did you know?

Coronary arteries on the surface of the heart supply materials to the network of capillaries in the heart muscles. Heart attacks are often caused by a coronary artery becoming blocked.

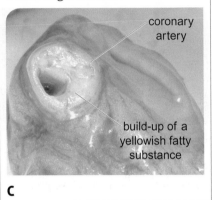

coronary artery

build-up of a yellowish fatty substance

C

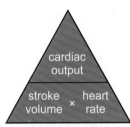

D This triangle can help you rearrange the equation for cardiac output. Cover up the quantity you want to calculate and write what you see on the right of the = sign.

Checkpoint

How confidently can you answer the Progression questions?

Strengthen

S1 Explain two ways in which the heart is adapted to its function.

Extend

E1 George has a leak in the valve between his right atrium and right ventricle. Suggest an explanation for why his breathing rate is fast.

SB8e Cellular respiration

Specification reference: B8.9; B8.10

Progression questions

- Why do organisms need to respire?
- Why is respiration an exothermic process?
- What are the differences between aerobic and anaerobic respiration?

A This is a jerboa. Small mammals lose heat more quickly than larger mammals and so have higher rates of respiration.

 1 Explain why your body needs a constant supply of energy.

 2 Explain why cellular respiration helps keep your body warm.

Your body requires a constant supply of energy for:

- moving
- keeping warm
- producing and breaking down substances.

Cellular respiration is a series of chemical reactions that release energy from **glucose**. Some energy is transferred out of the cells by heating, which helps keep many animals warm. Respiration is therefore **exothermic** (a process in which energy transfer increases the temperature of the surroundings).

The main type of cellular respiration is **aerobic respiration**, which needs oxygen. Most of the reactions in this process occur in the **mitochondria** of cells and can be summarised in this word equation:

glucose + oxygen → carbon dioxide + water

The circulatory system makes sure that cells have a good supply of oxygen (taken in by the lungs) and glucose (taken in by the small intestine). It also ensures that wastes are carried away from cells.

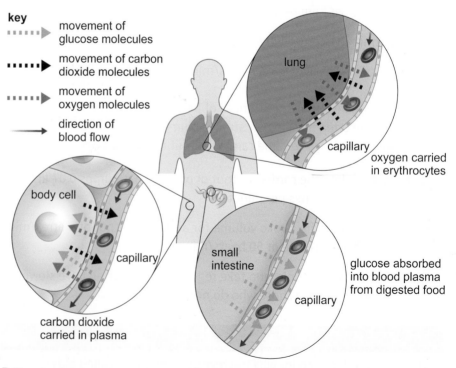

key
- movement of glucose molecules
- movement of carbon dioxide molecules
- movement of oxygen molecules
- direction of blood flow

lung

capillary

oxygen carried in erythrocytes

body cell

capillary

carbon dioxide carried in plasma

small intestine

capillary

glucose absorbed into blood plasma from digested food

B The circulatory system transports the reactants for respiration to cells and carries away waste products.

 3 Hummingbird muscle uses twice the amount of oxygen as mammal muscle. Describe a feature you would expect to find in hummingbird muscle cells.

 4 Which substance in the equation for aerobic respiration stores the most energy?

170

Exercise and anaerobic respiration

During exercise your muscles need more energy. The rate of aerobic respiration increases and your muscle cells take more oxygen and glucose from the blood. Your heart beats faster to get more blood to your muscle cells. You breathe faster and deeper to increase the amount of oxygen diffusing into the blood in your lungs. Faster breathing also allows your lungs to excrete more carbon dioxide.

During very strenuous exercise, oxygen is used up faster than it is replaced. When this happens, the amount of **anaerobic respiration** occurring in the cytoplasm of cells greatly increases. This form of cellular respiration does not require oxygen and produces **lactic acid**:

glucose → lactic acid

 5 Explain why your muscles need more energy when you exercise.

Did you know?

There are different types of anaerobic respiration in different organisms. The holes in some cheeses are caused by gases released by a form of anaerobic respiration that occurs in some bacteria. The ethanol in alcoholic drinks is produced by yeast using another form of anaerobic respiration.

D

Anaerobic respiration releases less energy from glucose than aerobic respiration. In animals, it also causes muscles to tire quickly. However, anaerobic respiration can release bursts of energy without needing a sudden increase in oxygen supply. This is important for animals that may need to move fast and suddenly, such as when sprinting away from a predator.

Heart and breathing rates can remain high after exercise, because extra oxygen is needed to replace the oxygen lost from blood and muscles. Extra oxygen is also needed to release the extra energy required to get rid of lactic acid.

 7 What are the waste product(s) of anaerobic respiration?

 8 a Explain why people continue to have high pulse rates after strenuous exercise.

 b Suggest why a fit person's pulse rate decreases more quickly after exercise than an unfit person's.

Exam-style question

State an advantage of anaerobic respiration for humans. *(1 mark)*

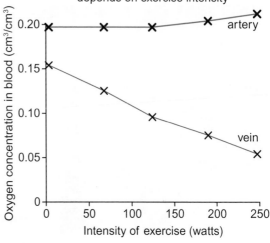

How the concentration of blood oxygen depends on exercise intensity

C Increasing exercise intensity affects the concentrations of gases dissolved in blood. Adapted from Carbon dioxide pressure-concentration relationship in arterial and mixed venous blood during exercise, Journal of Applied Physiology, Vol. 90, no. 5 (Sun, X-G, Hansen, J.E., Stringer, W.W., Ting, H., Wasserman, K.).

6 Look at graph C.

 a Suggest why the amount of oxygen in the artery increases as exercise intensity increases.

 b Explain why the line for the vein slopes downwards.

Checkpoint

How confidently can you answer the Progression questions?

Strengthen

S1 Design a table to compare and contrast aerobic and anaerobic respiration.

Extend

E1 Explain why small mammals lose heat more quickly than larger mammals.

E2 Explain how this affects their rate of respiration.

SB8e Core practical – Respiration rates

Specification reference: B8.11

Aim

Investigate the rate of respiration in living organisms.

A astronaut Kathryn Hire using a respirometer as part of an experiment in space

Many experiments are being done to find out the effects of space travel on human respiration. These experiments allow scientists to predict possible problems that astronauts might have while on space missions and to work out how much oxygen they will need. Scientists measure respiration rates using a respirometer, which measures the amount of oxygen used, the amount of carbon dioxide produced, or both.

Your task

You are going to use a simple respirometer to measure the oxygen consumption of some small organisms (e.g. mealworms) and to find out how the rate of respiration depends on temperature.

Method

Wear eye protection.

A Collect a tube with some soda lime, held in place with cotton wool. The soda lime absorbs carbon dioxide. Soda lime is corrosive, so do not handle it. The cotton wool is there to protect you and the organisms.

B Carefully collect some of the small organisms in a weighing boat.

C Gently shake the organisms out of the container and into the tube.

D Insert the bung and capillary tube, as shown in diagram B.

E Set up a control tube.

F Place both tubes into a rack in a water bath at a set temperature. It is best to tilt the rack slightly so that the capillary tubes hang over the side of the water bath at an angle.

G Wait for five minutes to let the organisms adjust to the temperature of the water bath.

H Hold a beaker of coloured liquid to the ends of the capillary tubes, so that liquid enters.

I Mark the position of the coloured liquid in the tube and time for five minutes.

J Mark the position of the coloured liquid again, and measure the distance it has travelled.

K Repeat the experiment at different temperatures.

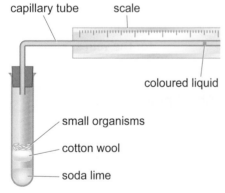

capillary tube   scale

coloured liquid

small organisms

cotton wool

soda lime

B a simple respirometer

Exam-style questions

1 a State the gas produced by aerobic respiration in the organisms.
   (1 mark)

   b Explain fully why the blob of coloured liquid moves in the capillary tube. *(3 marks)*

2 Describe one way in which the risk of harm is reduced in this experiment. *(1 mark)*

3 A student suggests using a small paintbrush to move the small organisms from a tray into her weighing boat. State why this would be a good idea. *(1 mark)*

4 a Describe how you would set up a control tube. *(1 mark)*

   b Explain why a control tube is necessary. *(2 marks)*

5 The experiment was set up using three large tubes at 25 °C. One tube was a control, one contained 20 g of active mealworms, and the other contained 20 g of slow-moving waxworms.

   a State the two control variables in this experiment. *(2 marks)*

   b State the independent variable. *(2 marks)*

   c In five minutes, the blob of coloured liquid moved 10 mm for the mealworms. Predict what would happen in the other two tubes. *(2 marks)*

   d Explain your predictions. *(2 marks)*

   e The moving of the liquid by 10 mm corresponds to a total change in volume of 5 mm³. Calculate the rate of respiration in terms of the volume of oxygen used up per gram of organism per minute. Show your working. *(2 marks)*

6 State the lowest and the highest temperature at which you would test the respiration rate in small organisms. Give reasons for your choices. *(2 marks)*

7 Table C shows the results of one experiment to measure the effect of temperature on the respiration rate of waxworms.

   a Explain why the measurements were repeated for each temperature. *(1 mark)*

   b Plot all the results on a scatter graph. *(2 marks)*

   c Identify the anomalous result. *(1 mark)*

   d Suggest an explanation for this anomalous result. *(1 mark)*

   e Draw a line of best fit through the remaining points. *(1 mark)*

   f Describe the correlation shown in your graph. *(1 mark)*

   g Suggest an explanation for this correlation. *(2 marks)*

Temperature (°C)	Distance moved by the blob in 5 min (mm)
10	9
10	9
10	10
15	12
15	15
15	13
20	17
20	20
20	18
25	25
25	25
25	28
30	10
30	33
30	38

C

The heart

Explain why a change in heart and breathing rates is an advantage when exercising strenuously. **(6 marks)**

. . . .

Student answer

Muscles need energy to make them work, and they need more energy when they work harder (like when you do exercise). Energy is released during respiration. As exercise gets harder, more and more energy is needed and so more and more respiration occurs. Aerobic respiration is the main type of respiration and needs oxygen and glucose:

glucose + oxygen → carbon dioxide + water [1]

The heart pumps faster to get more of the reactants for aerobic respiration to cells [2]. The breathing rate increases to get more oxygen into the blood [3]. A greater blood flow also helps to remove the carbon dioxide waste. If exercise is very strenuous, then anaerobic respiration increases. This uses up glucose but produces lactic acid, which is carried away faster by the increased flow of blood.

[1] This is an excellent start, clearly explaining what the body needs when exercising.

[2] This is a correct explanation for the first part of the question about heart rate.

[3] This is a correct explanation of the second part of the question about breathing rate. A small criticism here is that this sentence could have come first in this paragraph. Then all the information about the heart and increased blood flow would be together.

. . . .

Verdict

This is a strong answer. It clearly makes the link between respiration, exercise and the increase in breathing and heart rates. The answer correctly mentions both aerobic and anaerobic respiration and links them both to the increase in blood flow and breathing rates. The answer is well organised and uses scientific words correctly.

Exam tip

There are two things to consider in the question, 'heart rate' and 'breathing rate'. When you have finished writing your answer, ensure you look back at the question so you know you have covered everything.

Paper 2

SB9 Ecosystems and Material Cycles

Many flowers produce a scent that mimics the smell of a female insect. This attracts males of that insect species to visit the flower. During their visit, pollen from the flower attaches to the male. The pollen is taken to the next flower that the male visits. The orchid plant in the photo goes one step further by having flowers that mimic the shape and colour of a female digger wasp. Male wasps are so convinced by the deception, they try to mate with the flower. The orchid is dependent on the males of this one species of wasp to carry out their pollination.

The learning journey

Previously you will have learnt at KS3:

- how almost all life on Earth depends on photosynthesis in plants and algae
- about the interdependence of organisms, including food webs and insect pollination
- how organisms affect and are affected by their environment, including the accumulation of toxic materials.

In this unit you will learn:

- how ecosystems are organised
- how communities are affected by abiotic and biotic factors
- how the abundance and distribution of organisms are measured
- how energy is transferred through trophic levels
- about parasitism and mutualism
- how humans can affect ecosystems and the benefits of maintaining biodiversity
- about the importance of the carbon cycle, water cycle and nitrogen cycle
- how indicator species can be used to assess pollution levels
- why the rate of decomposition of food and compost can vary.

SB9a Ecosystems

Specification reference: B9.1; B9.3; B9.6

Progression questions

- What is a community of organisms?
- How are ecosystems structured?
- Why is interdependence in communities important?

A Plants interact with each other both above and below ground.

Organisms need **resources** to stay alive. Plants need space in which to get light, water, carbon dioxide, oxygen, warmth and mineral ions. Animals need oxygen, food and water. They may also need somewhere to shelter from the weather or avoid predation from other animals. This means that organisms are continually interacting with each other and with their environments.

 2 a State why plants have root systems.

1 Give reasons for each of the following.

    a Plants and animals need oxygen.

    b Plants need light and water.

    c Plants need mineral ions.

 b Use photo A to explain how plants interact with each other both above and below ground.

All the organisms and the environment in which they live form an **ecosystem**. An ecosystem may be large, such as a rainforest, or small, such as a pond.

All the organisms that live and interact in an ecosystem form a **community**. The community is made up of **populations** of different species. These species depend on each other for resources, so we say they are **interdependent**. Each population lives in a particular **habitat** within the ecosystem. A habitat includes the other organisms that affect the population and the local environment.

 3 a Sketch a diagram to show the relationship between the terms population, community and ecosystem.

 b Add an example of each term to your diagram, using the information from these pages.

B Coral reefs are the calcium carbonate outer skeletons built by tiny coral animals of different species. The reefs are also home to many other organisms, such as fish, starfish, sponges and turtles.

Abundance is a measure of how common something is in an area, such as its population size. Measuring population size by counting all the organisms in an area is often impossible. However, you can estimate population size by taking **samples** using a **quadrat**. Quadrats are placed randomly in the area, and the number of individuals in each quadrat is counted. The population size is estimated as:

$$\text{population size} = \text{number of organisms in all quadrats} \times \frac{\text{total size of area where organism lives}}{\text{total area of quadrats}}$$

A **food web** shows the feeding relationships between the organisms in a community. We can use a food web to help predict what will happen if there are changes in the ecosystem. For example, in diagram D, if all the herons die out, fewer frogs would be eaten. However, if the population of frogs increases, then more sticklebacks might be eaten.

C This quadrat has been placed on a school field. This sample includes two dandelion plants (long leaves) inside the large square. The square frame is divided into smaller squares to help count smaller plants, e.g. grass.

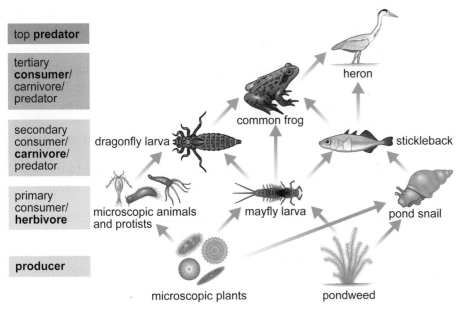

top **predator**

tertiary **consumer/** carnivore/ predator

secondary consumer/ **carnivore/** predator

primary consumer/ **herbivore**

producer

heron

common frog

dragonfly larva

stickleback

microscopic animals and protists

mayfly larva

pond snail

microscopic plants

pondweed

D a highly simplified food web for a pond ecosystem

4 In a $1\,\text{m}^2$ quadrat on a rocky shore there are 50 limpet shells. The total area of rocks is $450\,\text{m}^2$.

   a Estimate the total population size of limpets on the rocks.

   b Explain why it would be better to estimate the population size from a mean number of limpets from several randomly placed quadrats.

5 Use diagram D to answer these questions.

   a Name one herbivore in a pond ecosystem.

   b What is the top predator in this ecosystem?

   c What do frogs eat?

   d What is a producer?

6 Use diagram D to predict the effect on other organisms if all the pond snails died out. Explain your answer.

Checkpoint

How confidently can you answer the Progression questions?

Strengthen

S1 Define the terms: population, community, ecosystem, interdependence.

Extend

E1 Explain why diagram D is highly simplified and how this affects how correct your answer for question 6 might be.

Exam-style question

Describe two ways in which organisms in a community are interdependent.

(2 marks)

SB9b Energy transfer

Specification reference: B9.7B; B9.8B

Progression questions

- How is energy transferred from each trophic level, including in ways that are not useful to organisms?
- How does energy transfer limit the length of a food chain?
- How do you calculate the efficiency of energy transfer between trophic levels?

A Energy is measured in joules (J). The more red the colour, the greater number of joules of energy are transferred by light from the Sun to each square metre of the Earth's surface. Greener areas show less energy transfer by sunlight.

Did you know?

The efficiency of energy transfer in photosynthesis is around 95%, far higher than any mechanical systems we have made. The study of photosynthesis may lead to the development of bio-batteries as a much more efficient way of storing energy.

Map A shows the variation in the amount of energy from the Sun that reaches the Earth's surface. This variation affects both the physical conditions and how much life there is in each region.

Each year, over the whole Earth, photosynthesis captures over 3×10^{20} J of energy transferred by light from the Sun. Much of this energy is transferred to substances in new plant **biomass** (the mass of tissues). The rest is transferred to the environment by heating, during processes such as respiration. Other organisms cannot make use of energy transferred to the environment by heating and so these energy transfers are less useful for living things.

1 a State how energy is transferred to a plant.

b Explain why energy transferred to the environment by heating is less useful.

Diagram B shows what happens to the energy stored in plant biomass when it is transferred to a herbivore. The energy stored in the herbivore is then transferred to the carnivore that eats it, and so on through the **trophic levels** (feeding levels) of a food chain. We can think of this as energy flow through the **biotic** components of an ecosystem.

producer		primary consumer (a herbivore)		secondary consumer (a carnivore)
	energy flow			

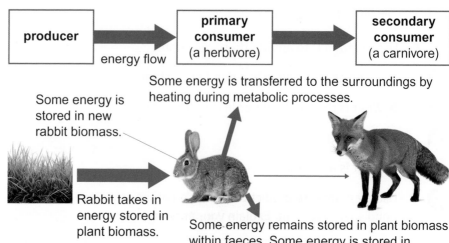

Some energy is transferred to the surroundings by heating during metabolic processes.

Some energy is stored in new rabbit biomass.

Rabbit takes in energy stored in plant biomass.

Some energy remains stored in plant biomass within faeces. Some energy is stored in substances in urine. This energy is not available to the rabbit's predator.

2 Identify the energy transfers to and from a herbivore.

3 Sketch a diagram to show the energy gain and losses in the fox in diagram B.

B energy flow through the trophic levels (producer, primary consumer, secondary consumer) of a simple food chain

We can show the energy transfers in an organism using a **Sankey diagram**, as in diagram C. This shows that, for every 25.0 kJ that the rabbit gains in its food, only 0.08 kJ is stored in new biomass. You can calculate the efficiency of energy transfer by the rabbit using this equation:

$$\frac{\text{energy transferred to biomass}}{\text{total energy supplied to organism}} = \frac{0.08}{25} = 0.0032$$

Efficiencies are usually given on a scale from 0 to 1, with 1 being complete efficiency. These values can be converted to percentages by multiplying by 100.

C a Sankey (energy transfer) diagram for a rabbit

25.0 kJ taken in as plant biomass

0.08 kJ stored in new biomass

12.52 kJ transferred to environment in faeces and urine

12.40 kJ transferred to environment by heating

If we measure the biomass of all the organisms at each trophic level in an ecosystem we can display them in a **pyramid of biomass**, as in diagram D.

 4 Use diagram C to calculate the percentage of energy transferred from the rabbit to the environment in a way that is no longer useful to organisms.

mass (g/m²) | trophic level

6	tertiary consumers
36	secondary consumers
78	primary consumers
809	producers

D A pyramid of biomass for an ecosystem. The width of each bar indicates the amount of biomass in each trophic level.

This kind of diagram usually has a pyramid shape because energy is transferred from the food chain to the environment at each trophic level. With less energy available, less biomass can be produced. We can calculate the percentage transfer of biomass between trophic levels in a pyramid of biomass. For example, from producers to primary consumers in diagram D:

$$\frac{\text{percentage biomass transfer from}}{\text{producers to primary consumers}} = \frac{78}{809} \times 100\% = 9.6\%$$

There is a maximum length of food chain in an ecosystem. The pyramid shape helps to explain why there is this limit. The energy stored in the biomass of the top trophic level is too little to support another level.

 5 Calculate the percentage biomass transfer from primary consumers to secondary consumers in diagram D.

 6 a Identify the number of trophic levels in the longest food chain in diagram D on *SB9a Ecosystems*.

b Use the pyramid of biomass model to explain why there are no predators of herons in the food web.

Exam-style question

Explain why a pyramid of biomass often has the pyramid shape shown in diagram D. *(2 marks)*

Checkpoint

How confidently can you answer the Progression questions?

Strengthen

S1 Explain why the percentage transfer of energy from one trophic level to the next is always less than 100%.

Extend

E1 Use photo A to suggest a reason why food chains in regions near the North Pole and South Pole usually have fewer trophic levels than those near the Equator.

SB9c Abiotic factors and communities

Specification reference: B9.2; B9.6

Progression questions

- What are abiotic factors?
- How do natural abiotic factors affect communities?
- How can pollution affect communities?

A The brown areas in these fields show where plants have died due to waterlogged soils caused by floods.

The **distribution** of organisms is where they are found in an ecosystem. Distribution can be affected by physical and chemical factors, such as temperature, rainfall and substances in the soil. These non-living factors are called **abiotic factors**. The effect of abiotic factors on the distribution of organisms can be measured using a **belt transect**. Quadrats are placed along a line in a habitat, and the abundance of organisms is measured as well as the abiotic factors in each quadrat position. Changes in abundance can show which abiotic factor has the greatest effect on the organisms.

 1 Name two abiotic factors related to climate.

Each species of organism has certain **adaptations** that mean the organism is suited to particular conditions. If abiotic factors change, then the distribution of organisms may also change.

Few organisms can survive a **drought** (lack of water) for long. Most land plants cannot survive if their roots are under water for long. If the climate changes resulting in more flooding or more drought, then many species in different communities may die out.

B Many sea birds such as the puffin, and predator fish such as haddock, depend on sandeels for food. Rising sea temperatures in the North Sea have reduced the numbers of microscopic animals that sandeels feed on.

Temperature also affects the distribution of organisms. For example, polar bears are adapted to living in cold regions, while cacti are adapted to living in hot deserts. However, all organisms have adaptations that make them suited to life at particular temperatures. A long-term rise or fall in temperature in an ecosystem will change the distribution of some organisms and so affect the whole community.

2 Explain why flooding may affect a whole community, not just the plants.

3 a Sketch a food web involving sandeels in the North Sea.

b Use your food web to explain why rising sea temperatures are linked to fewer haddock for us to eat and decreased numbers of puffins.

Light is essential for plants and algae to grow. In the oceans, most algae can only get enough light within 30 m of the surface. On land, light is limited within forests. In dense forests, few plants can grow on the forest floor.

 4 Suggest an explanation for where you would expect to find herbivores in the ocean.

 5 Explain why tree seedlings in a dense forest can start growing rapidly only when a mature tree has fallen.

Substances that cause harm in the environment are **pollutants** and cause **pollution**. Many human activities release pollutants. These can poison organisms or cause harm to organisms in other ways (such as plastics being eaten by fish and other organisms).

D In 2015 a dam collapsed at an iron mine on the River Doce in Brazil, releasing polluted muddy water into the river. High levels of poisonous mercury and arsenic killed fish and other river organisms.

C In dense forests, tree seedlings can start growing to full height only when a mature tree has fallen and allowed light to reach the forest floor.

Did you know?

The phrase 'mad as a hatter' refers to people who made hats in the 18th and 19th centuries. Mercury was used in the process. Long-term exposure to mercury damages the nervous system as well as many organs.

 6 State what is meant by pollution.

 7 Suggest how pollution from the collapsed dam in Brazil (see photo D) might have affected the whole community of organisms living beside the river.

Exam-style question

Explain why drought in an ecosystem can have long-term effects on the animals in a community. *(3 marks)*

Checkpoint

How confidently can you answer the Progression questions?

Strengthen

S1 A woodland is thinned by removing half the trees. Explain what effect this may have on low-growing plants in the following year.

Extend

E1 Plants that live in polar communities are generally much smaller than those living in tropical communities. Suggest an explanation for this.

SB9d Biotic factors and communities

Specification reference: B9.2

Progression questions

- What are biotic factors?
- How can competition affect communities?
- How can predation affect communities?

A Grey wolves hunt together to bring down elk.

Biotic factors are the organisms in an ecosystem that affect other living organisms. Yellowstone National Park is a huge area of protected land in northern USA. The Yellowstone ecosystem had included grey wolves until 1926, when they became extinct in the area due to hunting. The wolves had been the top predator in the community.

1 What is meant by a top predator?

2 Explain why Yellowstone Park is an ecosystem.

After 1926, the number of elk in the park increased rapidly. Their huge numbers caused overgrazing of many tree species. This left little food for other large herbivores, including beavers. Elk and beavers **compete** for food from trees. The numbers of coyotes (a kind of wild dog) also increased, because there was less **competition** from wolves for food such as young elk.

In 1995, grey wolves were reintroduced to Yellowstone Park. The aim was to increase **predation** of elk, whose numbers were out of control.

3 Use the information on this page to give two examples of competition, naming the resource that is being competed for.

4 Explain how lack of competition from wolves could lead to an increase in the number of coyotes.

B Wolves are larger than coyotes and chase them away, making it more difficult for coyotes to get food.

In large communities such as Yellowstone, many biotic factors may affect predator and prey numbers. However, in small communities, the numbers of a predator and its prey may be closely related in a **predator–prey cycle**. Graph C shows an example.

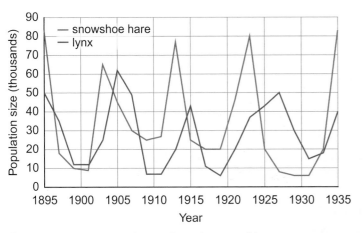

C Snowshoe hares are almost the only prey of lynxes in some areas of northern Canada. Their numbers are correlated in a predator–prey cycle.

The reintroduction of wolves to Yellowstone rapidly reduced the numbers of elk. This led to an increase in beaver numbers. Beavers change their surroundings by building dams, creating large pools and muddy areas. These new habitats allowed new species to grow in the area, increasing the **biodiversity** (the number of different species).

D Pools created by beaver dams allow the growth of species adapted to boggy areas, such as willow trees.

 6 Explain how the reintroduction of wolves to Yellowstone changed abiotic and biotic factors in that ecosystem.

 7 Explain why biodiversity in Yellowstone Park increased after wolves were reintroduced.

Exam-style question

Describe how introducing a new predator can affect a community through predation and competition. *(2 marks)*

5 Look at graph C.

 a Describe how the hare and lynx numbers are correlated.

 b Suggest an explanation for this correlation.

Did you know?

In the 19th century, humans hunted sea otters almost to extinction off western North America. This led to an increase in sea urchins, which sea otters eat. As a result, huge areas of kelp (seaweed) that sheltered many fish were destroyed as they were eaten by the sea urchins.

Checkpoint

How confidently can you answer the Progression questions?

Strengthen

S1 Explain why elk and beavers compete with each other in Yellowstone Park.

Extend

E1 Sketch a copy of graph C to show the snowshoe hare line. Add a line to your graph to show the population size of the hare's food (lichen). Explain the shape of your line.

Progression questions

- **H** What are indicator species?
- **H** How can indicator species be used as evidence of pollution?
- **H** How useful are indicator species as evidence of pollution?

H

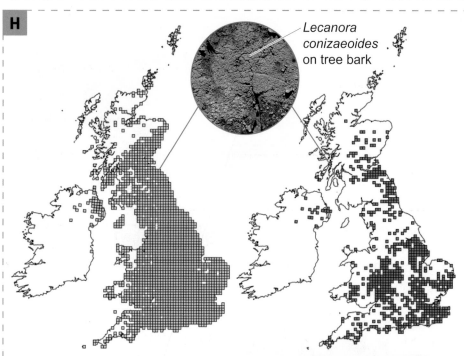

Lecanora conizaeoides on tree bark

A the distribution of *Lecanora conizaeoides* (left) in the 20th century and (right) in the 21st century

Air pollution

Lecanora conizaeoides is a species of **lichen** that grows on trees and buildings. Like all lichens, it is a mutualistic relationship between a fungus and an alga. Until about the last 20 years, it was the only lichen found in city centres and industrial areas. This is because it is the only lichen that can tolerate air polluted with sulfur-containing gases from burning fossil fuels. So, this lichen can be used as an **indicator species** for sulfur dioxide **pollution** in the air.

1 Describe the change in distribution of *Lecanora conizaeoides* shown in the maps in photo A.

Over the past 50 years, there has been a large fall in sulfur dioxide pollution in the air, and this has affected where *L. conizaeoides* is found. The distribution of other lichen species has also changed as the levels of other pollutants have changed. For example, increasing nitrogen oxide gases in the air, from vehicle exhausts, has increased the range of some lichen species and decreased the range of others.

2 Explain what is meant by an indicator species for pollution.

3 Explain why certain lichen species can be used as indicators of nitrogen oxide pollution.

Lichens are not the only indicators of air pollution. **Blackspot fungus** is a pathogen of roses. The fungus cannot grow well where there is a lot of sulfur pollution. So, roses growing in cities rarely suffered from blackspot infection.

4 a Explain why blackspot was not found on roses in cities before about 20 years ago.

b State whether photo B was taken in the middle of a city or in the countryside. Give a reason for your answer.

B Blackspot fungus infects roses, causing leaves to drop off the plant.

H Water pollution

Some **aquatic** (water-living) species of **invertebrates** are also useful pollution indicators. Water pollution can be caused by poisonous substances released by factories, such as mercury or detergents. Fertilisers and **sewage** are other common sources of water pollution. Substances in these (e.g. nitrates) cause **eutrophication**, which encourages the rapid growth of algae and plants. The bacteria that feed on the dead plants and algae then reduce the oxygen concentration in the water, which kills many animals. Different aquatic invertebrates are adapted to different concentrations of oxygen.

| indicator species | stonefly nymph | dragonfly nymph | freshwater shrimp | water louse | bloodworm | sludgeworm | no life |

| pollution level | low (stream) | low (pond) | slight | medium | high | extreme |

C Different aquatic invertebrates are found in water with different levels of pollution.

 5 State an aquatic invertebrate species that can be used as an indicator of clean water. Give a reason for your choice.

 6 Explain why sludgeworms and bloodworms are adapted to living in highly polluted water.

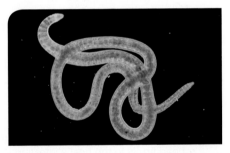

D Sludgeworms can live in highly polluted water because they contain haemoglobin. Bloodworms have the same adaptation.

Both water and air pollution can be measured using sensors. These give numerical data at the time of the measurement. Pollution indicator species do not give this level of detail in measurement, but they are useful as a simple assessment of the long-term health of an ecosystem.

 7 Give one advantage and one disadvantage of using living organisms as indicators of pollution.

Checkpoint

How confidently can you answer the Progression questions?

Strengthen

S1 Students sampling a stream in 2006 found bloodworms. Ten years later samples from the same stream included freshwater shrimps but no bloodworms. Suggest an explanation for this change.

Extend

E1 Jack's grandfather complains about blackspot on his roses and says he never had the problem 30 years ago. Suggest a conclusion that Jack can draw from this information, and what other evidence he needs to check that the conclusion is correct.

Exam-style question

Explain how blackspot fungus on roses can be used as an indicator of air pollution. *(2 marks)*

SB9f Parasitism and mutualism

Specification reference: B9.4

Progression questions

- How are some organisms dependent on other species?
- How does parasitism affect the survival of some organisms?
- How does mutualism help the survival of some organisms?

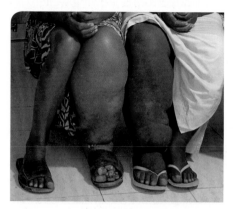

A Elephantiasis is caused by infection with roundworms. The roundworms absorb nutrients from human body fluids and become a problem when they block the flow of fluids in the body.

In most feeding relationships, a predator kills and eats its prey then moves on to find more prey. **Parasitism** is a different kind of feeding relationship in which one organism (the **parasite**) benefits by feeding off a **host** organism, causing harm to the host. The parasite lives in or on the host. The host may survive for a long time and continue to provide food for the parasite if the parasite causes limited harm.

1 Look at photo A.

 a Which is the parasite and which is the host organism?

 b Explain how the parasite benefits from its relationship with the host.

 c Explain how the host organism is harmed by its relationship with the parasite.

 2 In your own words, define the term parasite.

Some parasites, such as lice, live on the outside of their hosts. Others, such as tapeworms, live inside. All parasites have adaptations that help them survive in or on their host.

 3 How are lice adapted to feeding from their host?

 4 a Explain how a tapeworm could cause malnutrition in its host.

 b If malnutrition causes the tapeworm's host to die, explain why tapeworms do not all die out.

Hooks and suckers attach the worm's head firmly to the host's intestine wall.

Sharp mouthparts can pierce skin and suck blood.

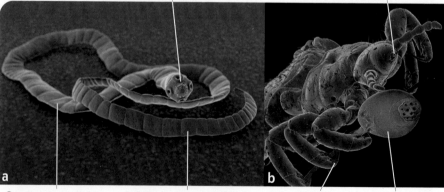

Segments contain male and female sex organs so fertilisation can occur.

A flattened body allows absorption of nutrients over whole surface without need for digestive or circulatory systems.

Sharp claws grip on to hair and skin.

Eggs are glued to hairs to prevent them falling off.

B Two examples of parasites. **a** Tapeworms are adapted to living inside their host's intestines. **b** Head lice are adapted to living on hair and skin.

Mutualism

Some organisms that live together both benefit from the relationship. These relationships are said to be **mutualistic**. For example, many flowers depend on insects for pollination. The flower benefits by being able to produce fertilised egg cells, and the insect benefits by collecting nectar or pollen from the flower, which it uses for food.

C A sea anemone's stinging tentacles protect clownfish from predators. Clownfish chase off the predators of the anemone and provide nutrients in their faeces, which help the anemone to grow.

 5 a How does the clownfish benefit from its relationship with a sea anemone?

 a How does the sea anemone benefit from its relationship with a clownfish?

Coral polyps form a special relationship with single-celled algae. The algae can live in the water surrounding corals, but are better protected inside a polyp. The algae photosynthesise and share the food they make with the coral animal.

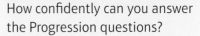

D Single-celled algae live inside the coral animals that build coral reefs.

 6 Describe how the relationship between a coral animal and algae is mutualistic.

 7 Explain the difference between parasitic and mutualistic relationships.

Checkpoint

How confidently can you answer the Progression questions?

Strengthen

S1 Using a suitable example, describe how a parasite is dependent on its host.

Extend

E1 One treatment for curing someone with elephantiasis is an antibiotic that kills *Wolbachia*. Explain why this works.

Exam-style question

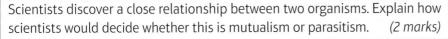

Scientists discover a close relationship between two organisms. Explain how scientists would decide whether this is mutualism or parasitism. *(2 marks)*

SB9g Biodiversity and humans

Specification reference: B9.9

Progression questions

- How does fish farming affect ecosystems?
- How does the introduction of new species affect biodiversity?
- How does eutrophication affect ecosystems?

A Salmon are farmed in pens, where they are better fed and grow faster than in the wild. They are also protected against predators and disease.

Many of the ways in which humans affect ecosystems can reduce biodiversity.

Fish farming

About 17 per cent of the protein eaten by people globally comes from fish. As the human population increases, we will need more fish. However, **overfishing** of wild fish stocks has damaged some aquatic (water) ecosystems. **Fish farming** aims to produce more fish and so reduce overfishing of wild fish.

 1 What is fish farming?

 2 Give two reasons why farming fish may be a better way to provide food for humans than catching wild fish.

 3 Describe two harmful effects that fish farming can have on an environment.

Fish farming causes problems because so many fish are kept in a relatively small space. Uneaten food, and faeces from the fish, sinks to the bottom of the water. This can change conditions, which may harm the wild organisms that live there. Parasites and disease spread more easily between fish in pens, so the fish need to be treated to keep them healthy.

Introducing species

Introducing new species to ecosystems can affect the **indigenous**, or **native**, species (organisms that have always been there). For example, sheep, cattle and soybeans are **native** to Asia but are farmed for food in many parts of the world where they are **non-indigenous**.

Did you know?

In the 2000s, sea lice from farmed salmon killed over 90 per cent of young wild salmon on one Canadian coast. Timing of treatment to kill the lice on farmed salmon was changed to just before the wild salmon passed through the area, reducing deaths of wild salmon to 4 per cent.

B Ring-necked parakeets are escaped pets that are now common in parts of the UK. Some smaller native birds are unable to compete for food with the parakeets.

Some species are introduced in order to affect an ecosystem, such as to reduce the numbers of another species that has got out of control. This often happens after humans have changed ecosystems and affected the food web. For example, cane toads from South America were introduced to Australia to control the numbers of cane beetles, which were eating sugar cane crops. Now the numbers of cane toads are a problem; the toads are poisonous and kill native animals.

native Australian frog

C Cane toads eat a wide range of indigenous species.

Eutrophication

Eutrophication is the addition of more nutrients to an ecosystem than it normally has. For example, this can happen when too much fertiliser is added to a field. Fertilisers help crop plants grow better, but will also increase the growth of other plants and algae. Diagram D shows how this can harm an ecosystem, causing a form of pollution.

 4 State three ways in which humans might introduce non-indigenous species into an ecosystem.

 5 Give two ways in which introduced species can affect a native food web.

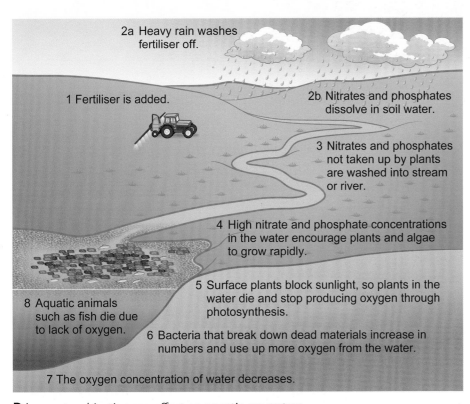

2a Heavy rain washes fertiliser off.

1 Fertiliser is added.

2b Nitrates and phosphates dissolve in soil water.

3 Nitrates and phosphates not taken up by plants are washed into stream or river.

4 High nitrate and phosphate concentrations in the water encourage plants and algae to grow rapidly.

5 Surface plants block sunlight, so plants in the water die and stop producing oxygen through photosynthesis.

6 Bacteria that break down dead materials increase in numbers and use up more oxygen from the water.

8 Aquatic animals such as fish die due to lack of oxygen.

7 The oxygen concentration of water decreases.

D how eutrophication can affect an aquatic ecosystem

 6 a How does adding fertiliser benefit a farmer's field?

 b Explain how eutrophication can change biodiversity in an aquatic ecosystem.

Did you know?

Lake Erie, a large lake on the border of Canada and the USA, became choked with algae in the 1960s and 70s due to fertiliser from fields and other nitrate sources. Strict pollution control has since reduced the problem.

Checkpoint

How confidently can you answer the Progression questions?

Strengthen

S1 Explain how too much fertiliser on a field can lead to the death of fish in a nearby river.

Extend

E1 American signal crayfish were introduced into the UK in the 1970s and bred for restaurants. Some escaped, and the wild population increased. Draw up a list of questions that need to be asked before a control programme is introduced to limit their numbers.

Exam-style question

Describe one benefit to biodiversity and one problem caused by farming fish rather than collecting them from the wild. *(2 marks)*

SB9h Preserving biodiversity

Specification reference: B9.10

Progression questions

- How can animal species be conserved?
- How can animal conservation protect biodiversity?
- How can reforestation affect biodiversity?

 1 What does reforestation mean?

 2 a Explain why the biodiversity of plants has increased in the Kielder area in the last 100 years.

 b Suggest a reason why increasing biodiversity of plants has affected the biodiversity of animals.

B Red squirrels are indigenous to the UK. Their numbers have decreased due to loss of their preferred conifer woodland habitat, and competition for food by introduced grey squirrels in broad leaved woodlands.

Northern England was once covered by forest. By 100 years ago, the trees had gone and it was mostly open moorland, where animals such as deer and grouse lived. Then a major **reforestation** project began. Kielder Forest was originally planted with conifer trees (e.g. pines) to provide wood. Today, both conifers and broad leaved trees are planted and some areas are left open, to increase the range of habitats and increase the number of species living in the area.

A The 250-square-mile Kielder Forest today is a mixture of habitats created by different tree species, moorland, grassland, a large artificial lake and many pools.

Kielder Forest is now home to species that are rare in other parts of the UK, including the osprey, goshawk and red squirrel. **Conservation** is when an effort is made to protect a rare or **endangered** species or habitat. For example, in Kielder, nesting platforms have been built in tall trees to encourage ospreys to nest. Also, any grey squirrels in the area are caught and killed, to help protect the red squirrels.

 3 Why are grey squirrels a threat to red squirrels?

 4 Explain how planting Kielder Forest has helped to conserve red squirrels.

Conservation of a species is easier if the species habitat is also protected. However, the habitats of many rare species are being damaged or destroyed. For example, tigers live in dense forests which are being cut down for wood and to create space for people to live. Also, people hunt tigers for fur and other body parts, and to reduce attacks on farm animals. Tigers are being bred in **captivity** (e.g. in zoos) to increase their numbers, but their habitats need to be rebuilt and protected too. These habitats also need to be linked so that tigers can roam more widely and find mates.

Preserving biodiversity is not just important for conserving individual species or communities. Areas with greater biodiversity can recover faster from natural disasters such as flooding. We also use plants and animals for food and as a source of medicines and other products. As conditions change, we may need new varieties of plants and animals to provide what we need. So it is important that we try to preserve as many species as we can.

C The South China tiger has not been seen in the wild for over 25 years and there are only about 70 in zoos.

 5 a Give two reasons why tigers need conservation.

 b Explain why tiger conservation is being carried out in captivity.

 c Suggest what else needs to be done before tigers can be returned to the wild.

 6 Give three reasons why preserving rainforests should be encouraged.

D Angkor Wat temple in Malaysia was abandoned in 1431. Even now, the forest that has regrown around it is not the same as the original indigenous forest, due to lack of seeds from some species in the soil.

Did you know?

Rainforests are incredibly biodiverse ecosystems. Rainforests contain about 500 species of tree per hectare (compared with about 6 to 12 per hectare in a UK forest). Many rainforest plants could be used to develop new medicines, and there are still many plants to discover.

 7 Suggest, with reasons, how the biodiversity of land cleared of rainforest could be increased more quickly than by leaving the area to recover naturally.

Exam-style question

Explain the benefits to wildlife of the reforestation of the Kielder area.

(2 marks)

Checkpoint

How confidently can you answer the Progression questions?

Strengthen

S1 How can planting an area of grassland with different kinds of trees increase biodiversity?

Extend

E1 A local landowner plans to develop a 5-hectare field. Write a letter outlining the advantages of replanting the area as mixed woodland.

SB9i Food security

Specification reference: B9.11B

Progression questions

- What is food security?
- Which factors affect food security?
- How is food security affected by different factors?

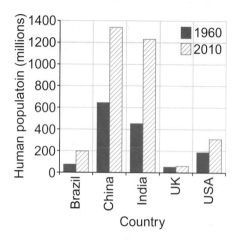

A population size in 1960 and 2010 for five countries

Food category	Supply (kg/person/year)	
	1961	**2011**
fish	9.01	18.93
meat	23.07	42.36
cereal	128.13	147.17

B world supply of different foods in 1961 and 2011

Food security means having access to enough safe and healthy food at all times. Improving agricultural methods, such as using fertilisers and more productive varieties and breeds, has increased food security in many places. However, the issue of food security is still difficult in very poor countries or where there is conflict. The issue of food security may also *become* more difficult due to the rising global human population. In 1960 there were just over 3 billion people; today there are over 7.5 billion.

 1 Suggest, with a reason, why food security can be difficult to achieve in very poor countries.

 2 Explain why the increasing human population could affect food security.

3 Suggest, with a reason, which country in graph A might have the greatest problem with food security in the next 10 years.

Table B shows how the supply of different foods is changing. Generally, as countries become wealthier, people prefer to eat more meat and fish. This affects the environment. For example, up to 15 times more protein is produced from soybean than from animals (meat) on the same area of land. Changing farming practices, such as eating farmed rather than wild fish (see *SB9g Biodiversity and humans*), can help. However, some people think we should replace meat and fish protein with more vegetable protein to help protect the environment.

4 a Comment on the percentage change in supply for each food category in table B between 1961 and 2011.

b Explain the risk to food security of changing food preferences as countries become wealthier.

C Could we protect the environment by becoming vegetarian?

Growing foods from crops also causes problems, particularly with **agricultural inputs** such as fertilisers. Fertilisers increase plant growth and the **yield** of crops. Over the past 50 years there has been a 700% increase in the amount of fertilisers used globally each year. Most of this fertiliser is made using chemical processes that need energy and release carbon dioxide. This raises concerns about **sustainability**. (If a process is sustainable, it is possible to continue it at the same level without causing harm, such as to the environment or to food security.)

5 Explain why some people think that the rate of increase in fertiliser use is not sustainable.

Increasing carbon emissions from many human activities are leading to **climate change**. This can lead to pests and pathogens moving into new areas. For example, midges, which are **vectors** for the virus that causes bluetongue disease, are killed by cold temperatures. Bluetongue disease was first seen in the UK in a cow, in 2007. Since then, it has spread to sheep and other cattle across the country.

One idea to reduce carbon emissions is to grow plants for **biofuels** to replace fossil fuels. The carbon released by burning a biofuel is only the amount removed from the air by the crop as it grew. In 2013, 0.8% of land in the UK that could have grown crops was used to grow biofuels, and this proportion is increasing. This raises other sustainability issues, of whether land should be used to grow food or fuel.

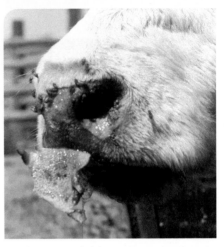

D Infection with bluetongue virus causes fever, swelling of the face and tongue and, in many cases, death.

 6 Explain why the spread of the bluetongue virus may be an effect of climate change.

E Elephant grass is a rapidly growing plant used as a biofuel. Here it is growing in a field that could have been used for growing food crops.

 7 Describe one advantage and one disadvantage of growing biofuels.

 8 Explain whether or not biofuels are more sustainable than fossil fuels.

Checkpoint

How confidently can you answer the Progression questions?

Strengthen

S1 Draw up a bullet point list to summarise threats to food security.

Extend

E1 You work for a government agency advising on farming policy. Draw up two recommendations for what should be done to improve food security in the UK for the next 50 years.

Exam-style question

Explain why food security could be affected by an increase in animal farming. *(3 marks)*

Progression questions

- Which materials cycle through ecosystems?
- How does water cycle through ecosystems?
- How is potable drinking water produced?

Living organisms need different substances from their environment to stay alive, such as water and carbon and nitrogen compounds. There are only limited amounts of these substances on Earth, so they must be recycled through organisms and the environment in order to support life.

Diagram A shows the **water cycle**. It illustrates how water moves through the abiotic parts of an ecosystem.

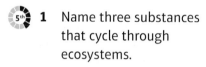

1 Name three substances that cycle through ecosystems.

2 The water cycle depends on the ability of water to change state with temperature. Identify the processes that cause water to change state in the water cycle.

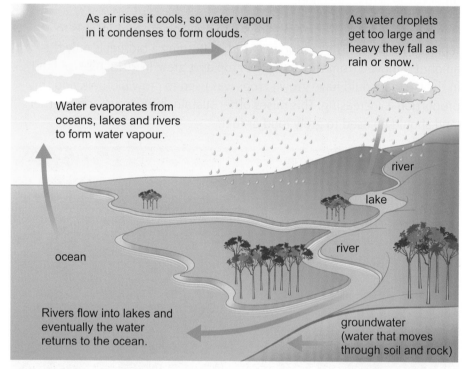

As air rises it cools, so water vapour in it condenses to form clouds.

As water droplets get too large and heavy they fall as rain or snow.

Water evaporates from oceans, lakes and rivers to form water vapour.

river

lake

ocean

river

Rivers flow into lakes and eventually the water returns to the ocean.

groundwater (water that moves through soil and rock)

A physical processes in the water cycle

3 Describe two ways in which we take water into our bodies.

4 Describe two ways in which water leaves our bodies.

5 Suggest how you could label diagram A to describe how water cycles through the biotic parts of an ecosystem.

6 Draw a flowchart to show how water from the environment is made potable.

Water makes up the majority of most organisms' body mass. For example, around 60 per cent of your body is water. Much of the cell cytoplasm is water, and reactions of substances often take place there. We are continually losing water to the environment, so we need to take in more water to replace it. Humans can only survive a few days without water.

There is plenty of fresh water in the environment, such as in rivers, lakes and underground. We can use this water for washing. To make it **potable** (safe for drinking), the water must be treated with chemicals and filtered, to remove dirt, pathogens and any toxic substances (such as some metal ions). The water may also be treated to improve the taste, by removing other non-toxic substances.

In places where there is drought, drinking water must either be collected from the air or extracted from sea water.

 7 Explain why water caught from clouds or mist is usually potable.

Obtaining fresh water from the sea or salty water is known as **desalination**. Several methods are used to do this, including **distillation** where the water is evaporated and then condensed and collected.

collection pipe at base of net

B Nets are used to catch droplets of water from clouds or mist in some desert areas. As long as equipment is clean, the water needs no treatment to make it potable.

 8 Explain how distillation can produce potable water from dirty water.

 9 Explain why desalination is an important source of potable water in Saudi Arabia.

Dirty water is put into the still here.

Water evaporates as it gets hotter inside the still.

Clean water is collected from the still here.

Water condenses under the cover and trickles to the bottom of the slope.

C A small solar still can supply fresh drinking water from dirty water where it is hot and sunny.

Water for desalination is taken from the nearby sea.

drinking water storage tanks

D Saudi Arabia is mostly desert. Over half the country's drinking water comes from desalination plants like this one at Shoaiba, though it requires a lot of energy for the process.

Exam-style question

Describe how water is cycled in the water cycle. *(3 marks)*

Checkpoint

How confidently can you answer the Progression questions?

Strengthen

S1 Explain why water is treated in the UK to make it suitable for drinking.

Extend

E1 Describe an advantage and a disadvantage of producing drinking water by desalination.

SB9k The carbon cycle

Specification reference: B9.13

Progression questions

- What is a decomposer?
- How is carbon cycled through an ecosystem?
- What is the role of decomposers in the carbon cycle?

Did you know?

We are made of stardust. All the carbon on Earth was originally formed in supergiant stars that exploded, scattering stardust across space. Carbon is the second-most abundant substance in the human body after water.

Pilobolus, the 'dung cannon' fungus, grows inside cow **faeces**, digesting the carbon compounds. The fungus produces capsules that contain spores (tiny new fungi). The capsule explodes off the top of the stalk onto fresh grass. When the cow eats the grass, the fungus is not digested and grows on the faeces when it leaves the cow. The fungus plays a key role in the **carbon cycle**.

1 cm

A The stalks of *Pilobolus* are less than 4 cm high, but can throw their capsules over 2 m away.

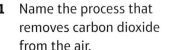 **1** Name the process that removes carbon dioxide from the air.

Carbon dioxide molecules in the atmosphere diffuse into plant leaves. Inside a leaf, photosynthesis may cause the carbon atom in the molecule to become part of another carbon compound, called glucose.

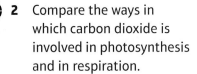 **2** Compare the ways in which carbon dioxide is involved in photosynthesis and in respiration.

If glucose is used by the plant for respiration, the carbon atom will become part of carbon dioxide again and be released back into the atmosphere. Alternatively, the glucose may be changed into other carbon compounds and used to make more plant **biomass**.

Carbohydrates, fats and proteins in plants all contain carbon atoms. When an animal eats a plant, some of these compounds are digested and taken into its body. The rest will leave the animal's body in faeces.

 3 Draw a labelled diagram to show what happens to the carbon in an animal that is eaten by a predator.

Some of the absorbed carbon compounds are used for respiration and some form waste products that are excreted in urine. The rest are used to build more complex compounds in the animal's tissue, making more animal biomass. If the animal is eaten by a predator, the same process happens again.

If plants and animals are not eaten and just die, their bodies are broken down by **decay**. Decay is caused by microorganisms that we call **decomposers**. Decomposers include fungi and bacteria, which also break down the carbon compounds in animal waste (e.g. faeces and urine).

Decomposers use some of the carbon compounds they absorb for respiration and to make more complex compounds in their cells. When they die, they will be decayed by other decomposers. If many large dead plants are buried so quickly that decomposers cannot feed on them, then over millions of years they may be changed into peat or coal by heat and pressure from the Earth. In the same way, oil and natural gas are formed from dead microscopic sea plants and animals.

Coal, peat, oil and natural gas are **fossil fuels**, as they contain carbon compounds that were in living organisms millions of years ago. Burning fossil fuels releases the carbon back into the atmosphere as carbon dioxide.

The movement of carbon through the biotic and abiotic components of the environment is called the **carbon cycle**.

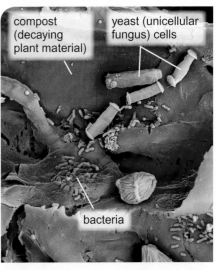

B Decomposers release enzymes into their surroundings to digest complex carbon compounds into smaller molecules that they can absorb. Magnification ×1500.

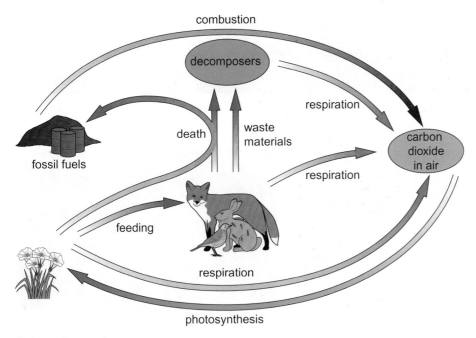

C the carbon cycle

 4 Which processes in the carbon cycle release carbon dioxide into the air?

 5 The natural carbon cycle does not include combustion of fossil fuels. Explain how the natural cycle usually keeps the amount of carbon dioxide in the air fairly constant.

Checkpoint

How confidently can you answer the Progression questions?

Strengthen

S1 Explain why *Pilobolus* (photo A) is important in the carbon cycle.

Extend

E1 Large areas of the Amazon rainforest have been cut down or burnt and replaced with grassland for cattle. Suggest what effect this has had on the local and global carbon cycle.

Exam-style question

What is the role of decomposers in the carbon cycle?   *(2 marks)*

SB9l The nitrogen cycle

Specification reference: B9.15

Progression questions

- Why do plants need nitrates?
- How do farmers increase the amount of nitrates in the soil?
- What is the role of bacteria in the nitrogen cycle?

Nitrogen is an unreactive gas that makes up around 80 per cent of the atmosphere. However, around 3 per cent of your body is composed of nitrogen compounds.

A Energy released during lightning storms can cause unreactive nitrogen in the air to form reactive nitrogen compounds that are added to the soil when it rains.

 1 **a** Describe the effect on plant growth of a lack of nitrogen.

 b Explain why nitrogen deficiency has this effect.

 2 A farmer plants a crop in a field. Suggest a reason why the nitrate concentration of the soil will change during the growing season in that field.

 3 What is meant by soil fertility?

 4 Explain how digging manure into a field can increase soil fertility.

Nitrogen in plants

Plants contain nitrogen compounds in proteins and DNA. To grow well, plants need nitrogen to make more of these compounds. They cannot use unreactive nitrogen from the air. Instead they absorb nitrogen compounds such as **nitrates** that are dissolved in soil water.

B These plants were planted at the same time. The one on the left is deficient in (has too little) nitrogen.

Bacteria and nitrates

Soil fertility is maintained by decomposers such as bacteria in the soil. These organisms release nitrogen compounds together with carbon compounds when they decompose dead plants and animals and their wastes.

Farmers make use of this decay process when they add **manure** (which includes animal waste) to their fields. Farmers may also spread artificial fertilisers onto fields to increase the soil fertility. The nitrogen compounds in fertilisers are soluble and dissolve in soil water.

Did you know?

Some plants that live on nitrogen-poor soils have adaptations that help them get their nitrogen from animals. Carnivorous plants include the sundew, which has sticky leaves, and the Venus flytrap, which has hinged leaves to catch insects.

Some soil bacteria can convert nitrogen gas into nitrogen compounds in the soil. They are called **nitrogen-fixing bacteria**. Some plants such as peas and beans have a mutualistic relationship with these bacteria. The bacteria are protected inside nodules in the plant roots, and the plant gets nitrogen compounds directly from the bacteria.

 5 Explain why the relationship between nitrogen-fixing bacteria and pea plants is described as mutualistic.

Farmers can also make use of this relationship to keep their soil fertile, by planting a crop of peas (or related plants) and then digging in the roots after the crop has been harvested. The following year, a different crop will benefit from the additional nitrogen compounds in the soil. Planting a sequence of crops in different years, such as wheat followed by potatoes followed by peas, is called **crop rotation**.

Diagram D shows how nitrogen cycles through the biotic and abiotic components of an ecosystem in the **nitrogen cycle**.

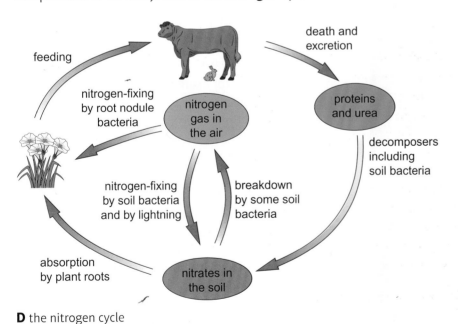

D the nitrogen cycle

C The root nodules of a pea plant contain nitrogen-fixing bacteria.

 6 Explain the purpose of crop rotation.

 7 Describe how the nitrogen compounds in your body come from nitrogen in the air.

Checkpoint

How confidently can you answer the Progression questions?

Strengthen

S1 Explain why farmers use manure, fertilisers and crop rotation in their fields.

Extend

E1 Until the early 1990s farmers used to burn the remains of crop plants after harvest and before planting a new crop. Now they usually plough in the remains. Explain why this change has helped to improve soil fertility.

Exam-style question

Explain how bacteria help make nitrates available for plant growth.

(2 marks)

SB9m Rates of decomposition

Specification reference: B9.17B; B9.18B; B9.19B

Progression questions

- How can the rate of decomposition of food be reduced?
- How can the rate of decomposition in composting be increased?
- How can the rate of decay be calculated?

A This body must have been covered with ice soon after death. The cold preserved soft tissues from the effects of decomposers.

vegetables in oil

fruit in sugar syrup

jam (fruit and sugar cooked to remove water)

vegetables in vinegar (dilute ethanoic acid)

dried salted meat

B Before there were fridges and freezers, food was stored in cool larders. Various other methods of preserving food were also developed.

In 1991, two walkers in the Alps, on the border of Italy, discovered a dead body in the ice. The skin and hair of the body were well preserved, so it was a shock to discover the body was over 5300 years old.

Food preservation

Usually the soft tissues of organisms are decayed (broken down) soon after death, by **decomposers**. These microorganisms grow best in warm and moist conditions, and many of them need oxygen.

This decay can cause problems, particularly with keeping food fresh. Most methods of food **preservation** rely on:

- reducing temperature, for example in fridges and freezers
- reducing water content, for example by salting and then drying meat to make ham or salami
- **irradiation** of packaged foods to kill decomposers
- reducing oxygen, for example storing foods in oil. Foods that easily decay (e.g. salad leaves) are often packaged in an unreactive gas, such as nitrogen.

 1 Give examples of two groups of decomposers.

 2 Explain why food is decayed by decomposers.

 3 **a** Identify two foods in photo B that were preserved by removing water.

 b Explain why removing water reduces the rate of decay.

Did you know?

Jerky is made by drying strips of salted meat, often beef. The first evidence for jerky is from about 1550 CE in South America, where it was made by the Quechua people. They used it as a source of protein on long journeys as it will last for months without decaying.

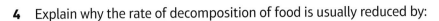

4 Explain why the rate of decomposition of food is usually reduced by:

 a reducing the temperature

 b reducing the oxygen concentration.

 5 Pickling reduces pH. Explain why this helps to preserve food.

Making compost

The effect of decomposers on dead plant material can be useful. Many gardeners collect waste garden material into a heap, and keep it until it is well-decayed, forming **compost**. The compost contains many of the nutrients that were in the plant tissues. However, the decay process leaves them in a form that makes it easier for plants to absorb. Spreading compost on a garden increases **soil fertility**. The way a compost heap is constructed affects the conditions inside the heap, and so affects the rate of compost formation.

 6 a Explain the features of the compost heap described in photo C.

 b Explain why insulating all round a compost heap may not be a good idea.

The rate of decay can be calculated from a quantity that changes over time. There are many quantities that could be measured, such as mass or area.

Spaces in the sides allow oxygen into the heap. The heap may be 'turned' after a few weeks, to get more oxygen into the centre.

A mixture of green and dead plant material, and a little soil, provides many different nutrients.

C Compost heaps are constructed to speed up decomposition. An insulated lid will be placed on top of this one when it is completed, to help keep it warm and moist.

Worked example

The mass of a fresh apple was 153 g. The apple was placed in a compost heap. Ten days later its mass was 37 g. Calculate the rate of decomposition of the apple, using the formula:

$$\text{rate of decomposition} = \frac{\text{mass lost}}{\text{number of days}}$$

mass lost = 153 − 37 = 116 g

$$\text{rate of decomposition} = \frac{116}{10} = 11.6 \text{ g/day}$$

7 The mass of a fresh apple was 142 g. The apple was placed in a compost heap in a cooler part of the garden than the one in the worked example. After 2 weeks its mass was 48 g.

 a Calculate the rate of decomposition of the 142 g apple using the formula in the worked example.

 b Explain any difference in the rate of decomposition between the two apples.

Exam-style question

Explain why keeping fresh meat in a fridge helps to preserve it for longer.

(2 marks)

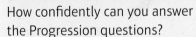

Checkpoint

How confidently can you answer the Progression questions?

Strengthen

S1 Sally wants to pack a lunch that will be safe to eat on the following day. Suggest what she could pack and how she should keep it until she eats it.

Extend

E1 Describe how you would build a compost heap to win a 'fastest ever compost' competition, and explain your design.

Using fertilisers

Over the past 50 years, the use of fertilisers in the world has increased from about 30 million tonnes each year to over 160 million tonnes each year. Discuss the use of fertilisers in agriculture.

(6 marks)

Student answer

Farmers use fertilisers on their crops because fertilisers add mineral ions to the soil, which plants need for healthy growth [1]. The benefit of using fertilisers is that crop yields are higher. This means that we have been able to grow more food over the past 50 years [2]. The human population of the world has been growing, so growing more food is necessary to avoid food shortages or starvation [3].

Using large amounts of fertilisers can cause problems [4]. They can be expensive for a farmer to buy. Fertilisers can also harm the environment if the mineral ions escape into surrounding water. This can cause eutrophication [5], leading to pollution that kills plants and animals living in the water.

[1] Starting with a description of what fertilisers do makes it easier to follow the arguments in the rest of the answer.

[2] This is a good example of linking ideas. The answer links why plants need fertilisers to what happens to crop yields when fertilisers are used, and then why we need more food.

[3] This last sentence gives a reason why fertiliser use has increased, and so a benefit for using fertilisers.

[4] The second paragraph clearly discusses problems using fertilisers.

[5] This is a good use of a scientific term.

Verdict

This is a strong answer. It covers both benefits and problems of fertilisers and explains why there has been an increase in the use of fertilisers. The answer has been arranged clearly so that each paragraph covers one part of the discussion, making it obvious that benefits and problems have both been explored.

The answer could have also talked about whether some of the problems caused by using fertilisers can be reduced, for example by crop rotation. The answer could have also included some information about the sustainable use of fertilisers.

Exam tip

'Discuss' questions expect you to explore as many aspects of the issue or argument as you can. Don't focus on one aspect of the question, for example why fertilisers cause pollution, but try to give an answer that covers lots of different points.

Glossary

ABO blood group System of sorting human blood into one of four phenotypes (A, B, AB, O) on the basis of antigens on blood cells.

abiotic factors Non-living conditions that can influence where plants or animals live e.g. temperature, the amount of light.

abundance A measure of how common something is.

acquired characteristic A characteristic of an organism that can change during its life due to a change in the environment.

acquired immune deficiency syndrome (AIDS) When a person's immune system has been damaged by HIV, so they are more likely to get secondary infections.

acrosome A cap-like structure on the head of a sperm cell that contains enzymes used to penetrate an egg cell.

activated To make active, such as when a lymphocyte is triggered by a pathogen to start dividing rapidly.

active site The space in an enzyme where the substrate fits during an enzyme-catalysed reaction.

active transport The pumping of particles across a cell membrane (usually from a region of lower concentration to a region of higher concentration, against the concentration gradient). This process requires energy.

adaptation The features of something that enable it to do a certain function (job).

adenine One of the four bases found in DNA. Often written as A. It pairs up with thymine.

ADH Antidiuretic hormone. Hormone produced by the pituitary gland that increases the permeability to water of the collecting duct in a nephron.

adrenal gland A gland located on top of a kidney that produces the hormone adrenalin. It can be referred to as 'an adrenal'.

adrenalin A hormone that is released from the adrenal gland when you are nervous or excited.

adult stem cell A stem cell found in specialised tissue that can produce more of the specialised cells in that tissue for growth and repair.

aerobic respiration A type of respiration in which oxygen is used to release energy from substances such as glucose.

agricultural input Something needed for growing food, such as farm equipment, fertilisers and pesticides for crops.

allele Most genes come in different versions called alleles.

alveoli Small pockets in the lungs in which gases are exchanged between the air and the blood.

anaerobic respiration A type of respiration that does not need oxygen.

anaphase A stage of mitosis in which the separated chromosomes move away from each other

ancestor An organism from which more recent organisms are descended.

antibiotic Medicine that helps people recover from a bacterial infection by killing the pathogen.

antibody A protein produced by lymphocytes. It attaches to a specific antigen on a microorganism and helps to destroy or neutralise it.

antigen A protein on the surface of a cell. White blood cells are able to recognise pathogens because of their antigens.

aorta A major artery leading away from the heart.

aquatic Living in water.

Ardi The nickname for a 4.4 million year old fossilised specimen of *Ardipithecus ramidus* that was discovered in Ethiopia.

artery A blood vessel that carries blood away from the heart.

artificial selection When people choose organisms with certain characteristics and use only those ones for breeding.

aseptic techniques Techniques used to keep out unwanted microorganisms, such as when growing microorganism cultures.

asexual reproduction Producing new organisms from one parent only. These organisms are genetically identical to the parent.

Assisted Reproductive Technology (ART) Technology that helps increase the chance of pregnancy, such as the use of hormones to stimulate egg release.

atrium An upper chamber in the heart that receives blood from the veins.

autoclave Machine used to sterilise equipment and culture media using pressure and heat.

auxins A group of plant hormones that affect the growth and elongations of cells.

axon terminal The small 'button' at the end of the branches that leave an axon.

axon The long extension of a neurone that carries an impulse away from the dendron or dendrites towards other neurones.

bacterial lawn plate A nutrient agar plate covered in a thin film of bacteria.

base (in DNA) A substance that helps make up DNA. There are four bases in DNA, often shown by the letters A, C, G and T.

belt transect A line in an environment along which samples are taken to measure the effect of an abiotic factor on the distribution of organisms.

Benedict's solution A solution used to detect the presence of reducing sugars (such as glucose) in foods.

binomial system The system of naming organisms using two Latin words.

biodiversity The variety of species in an area.

biofuel Fuel produced from biomass.

biological catalyst A substance found in living organisms that speeds up reactions i.e. an enzyme.

biological control Using organisms to kill problem organisms, such as pests or weeds.

biomass The total mass in living organisms, usually shown as the mass after drying.

biotic factors Living components (the organisms) in an ecosystem.

biuret test A test that uses copper sulfate solution and potassium hydroxide solution to test for proteins. The blue of the copper sulfate solution turns purple in the presence of proteins.

blackspot fungus Pathogen of roses that is killed by sulfur dioxide air pollution.

blood The fluid that carries oxygen and other substances from the heart to the body.

blood–brain barrier A natural filter that only allows certain substances to get from the blood into the brain (mainly due to cells in the capillary walls in the brain fitting together very closely).

body mass index (BMI) An estimate of how healthy a person's mass is for their height.

Bowman's capsule The start of a nephron where filtration occurs.

breed A group of animals of the same species that have characteristics that make them different to other members of the species

Bt toxin A natural insecticide made by the bacterium *Bacillus thuringiensis* that kills some kinds of caterpillar.

callus A clump of undifferentiated cells.

calorimeter Apparatus used to measure the energy content of substances by burning them and measuring temperature increase.

cancer A disease caused by the uncontrolled division of stem cells in a part of the body.

capillary A tiny blood vessel with thin walls to allow for the transfer of substances between the blood and tissues.

capsid The protein coat of a virus.

captivity Keeping something in unnatural surroundings, such as animals in a zoo.

carbon cycle A sequence of processes by which carbon moves from the atmosphere, through living and dead organisms, into sediments and into the atmosphere again.

cardiac output The volume of blood the heart can pump out in one minute. It is calculated using the equation cardiac output = stroke volume × heart rate.

cardiovascular disease A disease in which the heart or circulatory system does not function properly.

carnivore An animal that eats other animals.

carrier A person who has one allele for a genetic disorder and one 'healthy' allele. Carriers do not suffer from the disorder but their offspring may.

catalyst A substance that speeds up the rate of a reaction without itself being used up.

cataract Build up of protein in the lens of the eye.

cell cycle A sequence of growth and division that happens in cells. It includes interphase and mitosis, and leads to the production of two daughter cells that are identical to the parent cell.

cell (surface) membrane The membrane that controls what goes into and out of a cell.

cell sap The liquid found in the permanent vacuole in a plant cell.

cell wall A tough layer of material around some cells that is used for protection and support. It is stiff and made of cellulose in plant cells. Bacteria have a flexible cell wall.

cellular respiration Chemical processes by which living cells produce energy in the cell.

cellulose Plant cell walls are made of tough cellulose, which support the cell and allow it to keep its shape.

central nervous system (CNS) The main part of the nervous system that includes the brain and spinal cord.

cerebellum Part of the brain that controls balance, posture and fine muscle movements.

cerebral cortex The main part of the brain, which is used for most of our senses, language, memory, behaviour and consciousness.

cerebral hemispheres The cerebral cortex is divided into two cerebral hemispheres.

chalara dieback A disease of ash trees caused by a fungus called *Hymenoscyphus fraxineus*.

chamber An enclosed space. A human heart has four chambers.

chemical defence The use of chemical compounds by organisms to defend against attacks, such as lysozyme and hydrochloric acid in humans, and poisons and insect repellents in plants.

chemical reagents Substances that are used up in a chemical reaction.

chemotherapy Use of drugs to treat a disease, such as in treatment of cancer.

Chlamydia A bacterium that causes a sexually transmitted infection (STI).

chlorophyll The green substance found inside chloroplasts that traps energy transferred by light.

chloroplast A green disc containing chlorophyll found in plant cells. This is where the plant makes glucose through photosynthesis.

cholera A bacterial infection of the small intestine

chromosomal DNA The main bulk of DNA found in a cell. In humans, this DNA is found in chromosomes but the term is also used to describe the large loop of DNA found in bacteria.

chromosome A thread-like structure found in the nuclei of cells. Each chromosome contains one enormously long DNA molecule packed with proteins.

ciliary muscle A muscle that relaxes or contracts to change the shape of the lens in the eye.

ciliated (epithelial) cell	A cell that lines certain tubes in the body and has cilia on its surface.
cilium	A small hair-like structure on the surface of some cells. Plural is cilia.
circulatory system	The system that moves blood through the body. It consists of the heart, arteries, veins and capillaries.
cirrhosis	Damage to the liver caused by drinking large amounts of alcohol over a long period of time.
classification	The process of sorting organisms into groups based on their characteristics.
climate change	Change in weather patterns around the world.
clinical trial	The testing of a medicine on people.
clomifene therapy	A form of therapy used to stimulate ovulation.
clone	The offspring from asexual reproduction. All the cells in a clone are genetically identical to each other and to the parent's cells.
codominant	When two alleles for a gene both affect the phenotype, for example a person with the alleles for the A blood group and B blood group have blood group AB.
codon	A set of three bases (a triplet) found in DNA and RNA. The genetic code is formed from patterns of codons.
collecting duct	The final part of a nephron.
colour-blindness	An eye defect in which someone cannot see the full range of colours.
communicable disease	Any disease that can be spread directly from one person to another.
community	All the different organisms living and interacting with one another in a particular area.
companion cell	A specialised cell located in the phloem of plants.
competition	When organisms need the same resources as each other, they struggle against each other to get those resources. We say that they 'compete' for those things.
complementary base pair	Two DNA bases that fit into each other and link by hydrogen bonds.
compost	Waste vegetable material that has been decomposed, for use in increasing fertility of garden soil.
concentration gradient	The difference between two concentrations. There will be an overall movement of particles down a concentration gradient, from higher concentration to lower concentration.
concentration	The amount of a solute dissolved in a certain volume of solvent.
cone (cell)	A cell in the retina that detects different colours of light.
conifer	A type of tree that has needle-shaped leaves and has seeds contained in cones (not fruits).
conservation	The protection of an area or species to prevent damage.
constrict	To make narrower.
consumer	An animal that consumes (eats) other organisms.
continuous variation	Continuous data can take any value between two limits. Examples include length, mass and time. Continuous variation is when differences in a characteristic are continuous.
contraception	The prevention of pregnancy.
contract	To shorten.
converging lens	A lens that brings light rays together.
cornea	The transparent front part of the eye, which covers the iris and pupil.
corpus luteum	A structure that develops in an ovary after an egg cell has been released. It secretes progesterone.
correlation	A relationship between two variables, so that if one variable changes so does the other. This can be positive or negative.
crop rotation	Where a different crop is planted in the same field each year in a 3- or 4-year cycle, such as potatoes, oats, beans and cabbages. This helps to control the build-up of soil pests for each crop.
cross-sectional area (of a circle)	The area of a circle, calculated as πr², where r is the radius of the circle.
CT scan	A scan in which multiple X-rays are taken of part of the body and put together by a computer. CT stands for computed tomography.
cuticle	An outer covering that is not made of cells. Plant leaves have a cuticle covering the leaves.
cytokinesis	When the cytoplasm of a cell is separated as the cell membrane becomes pinched to form two daughter cells.
cytoplasm	The watery jelly inside a cell where the cell's activities take place.
cytosine	One of the four bases found in DNA. Often written as C. It pairs up with guanine.
daughter cell	A new cell produced from the division of a parent cell.
decay (biology)	A process in which complex substances in dead plant and animal biomass are broken down by decomposers into simpler substances.
deciduous	A word meaning 'shed at a certain time'. Deciduous plants shed their leaves in winter.
decomposer	An organism that feeds on dead material, causing decay.
deficiency disease	An illness due to insufficient supply of an essential dietary requirement.
dehydrated	Lacking in water.
denatured	A denatured enzyme is one where the shape of the active site has changed so much that its substrate no longer fits and the reaction can no longer happen.
dendrite	A fine extension from a neurone that carries impulses towards the cell body.
dendron	Large, long extension of a sensory neurone that carries impulses from dendrites towards the axon.

deoxygenated	Without oxygen.
dermis	Layer below the epidermis of the skin, which contains temperature receptors, sweat glands and erector muscles.
desalination	A process that produces fresh drinking water by separating the water from the salts in salty water.
diabetes	A disease in which the body cannot control blood glucose concentration at the correct level.
diagnosis	The identification of the cause of a problem.
dialysis	Process used to clean the blood of people with kidney failure. It involves the exchange of substances between blood and dialysis fluid across a partially permeable membrane.
diarrhoea	Loose or watery faeces
differentiation	The process by which a less specialised cell becomes more specialised for a particular function. The cell normally changes shape to achieve this.
diffusion	The random movement and spreading of particles. There is a net (overall) diffusion of particles from regions of high concentration to regions of lower concentration.
digestion	A process that breaks molecules into smaller, more soluble substances.
dilate	To make wider.
diploid	A cell or nucleus that has two sets of chromosomes. In humans, almost all cells except the sperm and egg cells are diploid.
direct proportion	A linear relationship in which the percentage change in a variable occurs with an equal percentage change in another variable. A direct proportion is seen as a straight line through the origin when the two variables are plotted on a graph.
discontinuous variation	Data values that can only have one of a set number of options are discontinuous. Examples include shoe size and blood group. Discontinuous variation is when differences in a characteristic are discontinuous.
disease	An illness that prevents the body functioning normally.
disease resistance	Unaffected or less affected by a certain disease.
distillation	The process of separating a liquid from a mixture by evaporating the liquid and then condensing it (so that it can be collected).
distribution	The pattern of places in which a certain organism can be found in an area.
distribution analysis	Looking at the pattern of where damaged plants occur, to help identify the cause of damage.
diverging lens	A lens that causes light rays to spread apart.
DNA	Deoxyribonucleic acid. A polymer made of deoxyribose sugar molecules and phosphate groups joined to bases.
DNA replication	When DNA molecules are copied before cell division occurs.
domain	The three main groups that organisms are now sorted into Archaea, Bacteria and Eukarya
dominant	Describes an allele that will always affect a phenotype as opposed to a recessive allele, whose effect will not be seen if a dominant allele is present.
dose	The correct quantity of a medicine that needs to be taken.
double helix	The shape of a DNA molecule, consisting of two helices.
drought	Lack of water.
drug	A chemical substance that alters the functioning of part of the body
ecosystem	An area in which all the living organisms and all the non-living physical factors form a stable relationship that needs no input from outside the area to remain stable.
effector	A muscle or gland in the body that performs an action when an impulse from the nervous system is received.
egg cell	The female gamete (sex cell).
egg follicle	Cells in the ovary that surround a developing egg cell. The follicle produces oestrogen.
elongation	When something gets longer such as a cell in a plant root or shoot before it differentiates into a specialised cell.
embryo	The ball of cells produced by cell division of the zygote. A very early stage in the development of a new individual.
embryonic stem cell	A cell from an early stage of division of an embryo that can produce almost any kind of differentiated cell.
endangered	An area or species that is at great risk of destruction.
endocrine gland	An organ that makes and releases hormones into the blood.
endothermic	A type of reaction in which energy from the surroundings is transferred to the products e.g. photosynthesis.
environmental variation	Differences between organisms caused by environmental factors such as the amount of heat, light and damage by other organisms. These differences are called acquired characteristics.
enzyme	A protein produced by living organisms that acts as a catalyst to speed up the rate of a reaction
epidemic	When many people over a large area are infected with the same pathogen at the same time.
epidermis	Outer layer of skin (or other external organ).
epidermis cells	Cells that form a surface layer of cells in an external plant or animal organ.
epithelial cell	A cell found on the surfaces of internal organs.
erector muscle	Muscle in the skin dermis that contracts and raises a body hair
erythrocyte	Another term for red blood cell.

ethanol emulsion test	A test using ethanol to detect lipids in foods.
ethene	A gaseous plant hormone that is involved in the ripening of fruit.
eukaryotic	A cell with a nucleus is eukaryotic. Organisms that have cells like this are also said to be eukaryotic organisms.
eutrophication	The addition of more nutrients to an ecosystem than it normally has.
evolution	A change in one or more characteristics of a population over a long period of time.
excrete	To expel waste materials that have been produced inside an organism.
exothermic	A type of reaction in which energy is transferred to the surroundings from the reactants e.g. combustion.
extinction	When a species dies out.
eyepiece lens	The part of the microscope you look down
faeces	Undigested food that forms a waste material.
family pedigree chart	A chart showing the phenotypes and sexes of several generations of the same family, to track how characteristics have been inherited.
fertilisation	Fusing of a male gamete with a female gamete.
fertiliser	Substances that add plant nutrients to soil, such as artificial fertilisers containing nitrogen or manure (a natural fertiliser made from animal waste).
fertility (of soil)	The nutrient content of a soil, which affects how well plants grow.
fever	A core body temperature that is too high (above 38 °C).
Fick's law	The relationship between the different variables that affect diffusion: rate of diffusion ∝ surface area × concentration difference / thickness of membrane
field of view	The circle of light you see looking down a microscope
fight-or-flight response	Several responses that prepare the body for sudden action, including increased heart rate, increased blood flow to muscles and the release of glucose into the blood.
filtration (in kidney)	Separating large molecules from smaller ones, as in the glomerulus and Bowman's capsule of a nephron.
first convoluted tubule	Part of a nephron where selective reabsorption of glucose and some mineral ions takes place.
fish farming	Growing fish in pens for food.
flagellum	A tail-like structure that rotates, allowing a unicellular organism to move. Plural is flagella.
fluid	A liquid or a gas.
food chain	A diagram that uses arrows to show the flow of energy through organisms that depend on each other for food.
food security	Having access to enough safe and healthy food at all times.
food web	A diagram of interlinked food chains. It shows how the feeding relationships in a community are interdependent.
fossil fuel	A fuel formed from the dead remains of organisms over millions of years e.g. coal, oil and natural gas
FSH	A hormone produced by the pituitary gland which causes egg cells to mature in the ovaries.
gamete	A haploid cell produced by meiosis used for sexual reproduction.
gamma ray	A high-frequency electromagnetic wave emitted from the nucleus of a radioactive atom. Gamma rays have the highest frequencies in the electromagnetic spectrum
gas exchange	A process in which one gas diffuses across a membrane and another gas diffuses in the opposite direction.
gene	A section of the long strand of DNA found in a chromosome that often contains instructions for a specific protein.
genetic code	A set of rules defining how the base order in DNA or RNA is turned into a specific sequence of amino acids joined in a polypeptide chain.
genetic diagram	A diagram showing how the alleles in two parents may form different combinations in the offspring when the parents reproduce.
genetic disorder	A disorder caused by faulty alleles.
genetic engineering	Altering the genome of an organism, usually by adding genes from another species. Also called genetic modification.
genetic variation	Differences between organisms caused by differences in the alleles they inherit from their parents, or differences in genes caused by mutation. Also called inherited variation.
genetically modified organism (GMO)	An organism that has had its genome artificially altered.
genome	All of the DNA in an organism. Each body cell contains a copy of the genome.
genotype	The alleles for a certain characteristic that are found in an organism.
genus	A classification group for closely-related species with similar characteristics. The genus name is the first word in the scientific name for a species (the second word is the 'species name').
gibberellins	A group of plant hormones that cause seeds to germinate, and flowers and fruits to form.
glomerulus	A network of blood capillaries associated with the Bowman's capsule of a nephron.
glucagon	A hormone that increases blood glucose concentration.
glucose	A sugar produced by the digestion of carbohydrates and needed for respiration.
glycogen	A polymer storage material made from glucose.
gravitropsim	A growth response to the stimulus of gravity.
growth	A permanent increase in the number and/or size of cells in an organism.
guanine	One of the four bases found in DNA. Often written as G. It pairs up with cytosine.
guard cell	A pair of guard cells open and close plant stomata.
haemoglobin	The red, iron-containing pigment found in red blood cells.
haemorrhagic fever	A disease which includes a fever (high body temperature) and internal bleeding, such as caused by the Ebola virus.
haploid	A cell or nucleus that has one set of chromosomes. Gametes are haploid.
health	A state of complete physical, social and mental well-being.
heart	A muscular organ in the circulatory system that pumps blood around the body.
heart attack	When the heart stops pumping properly due to a lack of oxygen reaching part of its muscle tissue.
heart rate	The number of heart beats in a unit of time, usually per minute (beats/min).
heart valve	Flaps of tissue between the atria and ventricles of the heart that stop blood flowing in the wrong direction when the heart muscle contracts.
herbivore	An animal that eats plants.
herd immunity	When the majority of people in a group are immunised, which provides protection to the few who are not by reducing their chances of meeting an infected person.
heterozygous	When both the alleles for a gene are different in an organism.
homeostasis	Controlling the internal environment of the body at stable levels.
homozygous	When both the alleles for a gene are the same in an organism.
hormonal system	The collection of glands in the body that release hormones.
hormones	Chemical messengers that are made in one part of the body and are carried in the blood to other parts, which they affect.
host	An individual that can be infected by a certain pathogen.
Human Genome Project	The international project that mapped the base pairs in the human genome.
human immunodeficiency virus (HIV)	A virus that attacks white blood cells in the human immune system, often leading to AIDS.
hybridoma cell	A cell made by fusing a lymphocyte and a cancer cell.
hydrogen bond	A weak force of attraction caused by differences in the electrical charge on different parts of different molecules.
hygiene	Keeping things clean by removing or killing pathogens
hypothalamus	Part of the brain that is important in monitoring and controlling body temperature.
hypothermia	A core body temperature that is too low (below 36 °C).
immune	When a person does not fall ill after infection because their immune system attacks and destroys the pathogen quickly.
immune system	All the organs in the body that protect against disease. It includes barriers, such as the skin, together with organs that help to kill pathogens.
immunisation	Making someone immune, for example by vaccinating them.
impulse	An electrical signal transmitted along a neurone.
index	A small raised number after a unit or another number to show you how many times to multiply that number together. For example, 10^3 means multiply 10 together 3 times ($10 \times 10 \times 10$).
indicator species	Organism whose presence indicates the presence or absence of certain types of pollution.
indigenous	Organisms that have always been in an area. Another word for native.
inhibit	To stop or slow down a process
insecticide	A substance used to kill insect pests.
insulin	A hormone that decreases blood glucose concentration. It is used in the treatment of type 1 diabetes.
interdependent	When organisms in an area need each other for resources e.g. for food and shelter.
interphase	The stage during which a cell prepares itself for cell division. DNA replication takes place and additional subcellular structures are produced.
inverse square law	A mathematical relationship in which a quantity varies in inverse proportion to the square of the distance from the source of the quantity.
inversely proportional	A relationship between two variables in which if one variable doubles, the other halves.
invertebrate	Animal without bones, such as an insect or worm.
iodine solution	Solution used to test for the presence of starch.
iris	The coloured part of the eye. Muscles in it control the diameter of the pupil.
irradiate	Exposing something to ionising radiation e.g. to kill decomposers in certain foods using gamma rays.
IVF	Fertilising an egg cell by placing it in a sterile container and then adding sperm cells.
kidney	The organ that removes urea, excess water and other substances from the blood to form urine.
kidney failure	When both kidneys don't work properly.
kingdom	There are five kingdoms into which organisms are usually divided: plants, animals, fungi, protists and prokaryotes.

lactic acid	The waste product of anaerobic respiration in animal cells.
lens (eye)	Part of the eye that further converges light rays (which have been converged by the cornea) to focus them on the retina.
lesion	An area of damage, such as the cracks in bark caused by chalara dieback fungus in ash trees.
LH	A hormone produced by the pituitary gland which causes ovulation.
lichen	A mutualistic relationship between a fungus and an alga. The presence of some species can indicate different levels of air pollution.
lifestyle	The way we live that can our affect bodies, such as what we eat, whether we smoke or do exercise.
ligase	An enzyme that joins two DNA molecules together.
lignin	A type of polymer that is combined with cellulose in some plant cells to make them woody, e.g. in xylem cells.
limiting factor	A single factor that when in short supply can limit the rate of a process, such as photosynthesis.
linear relationship	A relationship between two variables shown by a straight line on a graph.
lipid	A substance in a large group of compounds that includes fats and oils.
lock-and-key model	A model that describes the way an enzyme catalyses a reaction when the substrate fits within the active site of the enzyme.
long-sightedness	An eye condition in which close objects appear blurred.
loop of Henle	Long loop of a nephron involved in osmoregulation..
Lucy	The nickname for a 3.2 million-year-old fossilised specimen of *Australopithecus afarensis*.
lymphocyte	A type of white blood cell that produces antibodies
lysis	When the cell membrane of a cell breaks open, releasing everything inside the cell.
lysogenic pathway	The pathway in a virus life cycle where the virus genetic material inserts into the cell's genetic material and is replicated each time the cell divides.
lytic pathway	The pathway where a virus enters a cell, takes over the cell's replication process to produce more viruses, and causes lysis of the cell as the new viruses are released.
lysozyme	An enzyme produced in tears, saliva and mucus that damages pathogens.
magnification	The number of times larger an image is than the initial object that produced it.
malaria	A dangerous disease caused by a protist that causes serious fever, headaches and vomiting and can lead to death.
malnutrition	Health problems caused by a diet that contains too little or too much of one or more nutrients.
manure	A mixture containing animal waste that is added to soil to improve its fertility.
mean	An average calculated by adding up the values of a set of measurements and dividing by the number of measurements in the set.
median	The middle value in a data set
medulla oblongata	Part of the brain at the top of the spinal cord. It controls breathing and heart rate.
meiosis	A form of cell division in which one parent cell produces four haploid daughter cells.
memory lymphocyte	A lymphocyte that remains in the blood for a long time after an infection or vaccination.
menopause	When the menstrual cycle stops completely.
menstrual cycle	A monthly cycle involving the reproductive organs in women.
menstruation	The breakdown and loss of the lining of the uterus at the start of a woman's menstrual cycle.
meristem	A small area of undifferentiated cells in a plant where cells are dividing rapidly by mitosis.
meristem cell	A stem cell found in a plant meristem.
messenger RNA (mRNA)	A single strand of RNA produced in transcription.
metabolic rate	The overall rate at which chemical reactions occur within the body.
metabolism	All the chemical reactions that occur in your body.
metaphase	The stage of mitosis when the chromosomes line up across the middle of the cell.
microvillus	A tiny fold in the cell surface membrane of a cell. Microvilli (plural) are used to increase the surface area of a cell.
mineral ion	An ion from a naturally-occuring compound in the soil, which is important for plant growth.
mitochondrion	A sub-cellular structure (organelle) in the cytoplasm of eukaryotic cells where aerobic respiration occurs. Plural is mitochondria.
mitosis	The process of cells dividing to produce two diploid daughter cells that are genetically identical to the parent.
MMR	Stands for measles, mumps and rubella. The vaccine given to develop immunity to these diseases.
mode	The most common value in a data set.
monoclonal antibodies	Many identical antibodies.
monoculture	A large area of one kind of crop.
monohybrid inheritance	The study of how the alleles of just one gene are passed from parents to offspring.

monomer	A small molecule that can join with other molecules like itself to form a polymer.
motor neurone	A type of neurone that carries impulses to effectors.
mucus	A sticky substance secreted by cells that line many openings to the body
multicellular organism	An organism made up of many cells.
mutation	A change to a gene caused by a mistake in copying the DNA base pairs during cell division, or by the effects of radiation or certain chemicals.
mutualistic	A relationship between individuals of different species where both individuals benefit.
myelin sheath	Fatty covering around the axons of many neurones.
native	Another term for indigenous.
natural selection	A process in which certain organisms are more likely to survive and reproduce than other members of the same species because they possess certain genetic variations.
negative feedback	A control mechanism that reacts to a change in a condition, such as temperature, by trying to bring the condition back to a normal level.
nephron	Long tubule found in the kidney where filtration of blood occurs. Useful substances are reabsorbed leaving wastes and other substances that are not required in urine.
nerve cell	Another term for neurone.
nervous system	An organ system that contains the brain, spinal cord and nerves and carries impulses around the body. This system helps you to sense and respond quickly to changes inside and outside of your body.
neurone	A cell that transmits electrical impulses in the nervous system.
neurotransmission	Impulses passing from neurone to neurone.
neurotransmitter	A substance that diffuses across the gap between one neurone and the next at a synapse, and triggers an impulse to be generated.
nitrate	A compound that contains nitrogen in the form of a nitrate ion.
nitrogen cycle	A sequence of processes by which nitrogen moves from the atmosphere through living and dead organisms, into the soil and back to the atmosphere.
nitrogen-fixing bacteria	Bacteria that can take nitrogen from the atmosphere and convert it to more complex nitrogen compounds such as ammonia.
non-communicable	When a disease cannot be spread from animal to animal, or person to person.
non-indigenous	Organisms that have been introduced to an area where they haven't been before.
non-reducing sugars	Larger sugar molecules, such as sucrose, that do not react with Benedict's solution.
normal distribution	When many individuals have a middle value for a feature with fewer individuals having greater or lesser values. This sort of data forms a bell shape on charts and graphs.
nuclear pore	A small hole in the membrane around the nucleus.
nucleus	The 'control centre' of a eukaryotic cell.
nutrient agar	Agar containing nutrients. Used for growing cells, such as in bacterial lawn plates.
obesity	A condition in which someone is overweight for their height and has a BMI above 30.
objective lens	The part of the microscope that is closest to the specimen.
oestrogen	A hormone produced by the ovaries that is important in the menstrual cycle.
optic nerve	The nerve that transmits impulses from the retina to the brain.
optimum pH	The pH at which an enzyme's rate of reaction is greatest, or at which a population of microorganisms grows most rapidly.
optimum temperature	The temperature at which an enzyme's rate of reaction is greatest, or at which a population of microorganisms grows most rapidly.
organ donation	The transfer of a healthy organ (e.g. kidney) into the body of someone whose own organ has failed.
osmoregulation	The control of the balance of water and mineral ions in the body.
osmosis	The overall movement of solvent molecules in a solution across a partially permeable membrane, from a dilute solution to a more concentrated one.
ovary	The organ in the female reproductive system that releases egg cells and the hormones oestrogen and progesterone.
overfishing	Taking more fish from a population than are replaced by the fish reproducing so that the population falls over time.
oviduct	A tube that carries egg cells from the ovaries to the uterus in females. Fertilisation happens here.
ovulation	The release of an egg cell from an ovary.
oxygenated	With oxygen.
palisade cell	Tall, column-shaped cell near the upper surface of a plant leaf.
pancreas	An organ in the body that produces some digestive enzymes, as well as some hormones.
parasite	An organism that lives on or in a host organism and takes food from it while it is alive.
parasitism	A feeding relationship in which a parasite benefits and its host is harmed.
partially-permeable	Describes a membrane that will allow certain particles to pass through it but not others. Another term for semi-permeable.
passive	A process that does not require energy is passive.
pathogen	An organism (usually a microorganism) that causes disease.

penicillin	The first kind of antibiotic. It was extracted from a mould.
pentadactyl limb	A limb that has five digits (fingers and thumbs). Amphibians, reptiles, birds and mammals share this characteristic
percentile	The value of a variable below which a certain percentage of observations fall. For example, in an ordered set of data, the 20th percentile is the value at which 20% of the data points are the same or lower.
period	The 'bleed' that occurs during menstruation.
permeability	A measure of how well a membrane allows substances to pass through it. More permeable membranes allow more substances through.
pest	Animals that cause problems, such as by damaging crops.
PET scan	A scan in which a radioactive marker is used to pinpoint certain areas in the body (such as very active cells). PET stands for positron emission tomography.
PET scanner	A scanner used to identify the position of radioactive substances inside the body.
phagocyte	A white blood cell that is capable of engulfing microorganisms, such as bacteria.
phenotype	The characteristics produced by a certain set of alleles
phloem	Living tissue formed of sieve tubes and companion cells that transports sugars and other compounds around a plant.
photosynthesis	A series of enzyme-catalysed reactions carried out in the green parts of plants. Carbon dioxide and water combine to form glucose. This process requires energy transferred by light.
phototropism	Growth in response to the stimulus of light.
physical barrier	A barrier that makes it difficult for pathogens to get into the body, such as skin, mucus and cilia in animals, and cuticle and cell walls in plants.
pituitary gland	An organ just below the brain that releases many different hormones. It can be referred to as 'the pituitary'.
plant hormone	A substance released by certain cells in a plant that has an effect on other cells, usually causing the cells to grow and develop in a certain way.
plasma	The straw-coloured liquid component of blood.
plasmid	A small loop of DNA found in the cytoplasm of bacteria.
plasmid DNA	DNA found in plasmids.
platelet	Cell fragments that are important in the clotting mechanism of the blood.
pollutant	A substance that harms living organisms when released into the environment.
pollution	Harm caused to the environment, such as by adding poisonous substances or by abnormally high amounts of a substance in the air.
polymer	A long-chain molecule made by joining many smaller molecules (monomers) together.
polypeptide	A chain of amino acids.
population	A group of one species living in the same area.
potable	Suitable for drinking.
potometer	A device used for measuring the rate of water uptake by a plant.
pre-clinical	The testing of a drug before it is tried on humans, including testing on cells or tissues and on other animals.
precipitate	An insoluble substance that is formed when two soluble substances react together in solution.
predation	When one animal species kills and eats another animal species.
predator–prey cycle	The regular variation in numbers of predators and numbers of prey within a feeding relationship.
pregnancy	The time during which a fertilised egg develops in the uterus until the birth of the baby.
preservation	Keeping something from being damaged.
primary consumers	The first trophic level of consumers. Primary consumers are herbivores.
probability	The likelihood of an event happening. It can be shown as a fraction from 0 to 1, a decimal from 0 to 1 or as a percentage from 0% to 100%.
producer	An organism that makes its own food, such as a plant using photosynthesis. Producers form the first trophic level in a food web or food chain.
product	A substance formed in a reaction
progesterone	One of the hormones released by the ovaries.
prokaryotic	A cell with no nucleus is prokaryotic. Organisms with cells like this are said to be prokaryotic organisms, e.g. bacteria.
prophase	The stage of mitosis in which the nucleus starts to break down and spindle fibres appear.
protein	A polymer made up of amino acids.
protist	An organism that belongs to a kingdom of eukaryotic and mainly single-celled organisms (also called a 'protoctist').
puberty	The stage of life when the body develops in ways that allow reproduction, e.g. the production of sperm cells in testes and the release of egg cells from ovaries.
pulmonary artery	The artery that carries blood from the right ventricle to the lungs.
pulmonary vein	A vein that carries oxygenated blood from the lungs to the left atrium.
pulse	A shock wave that travels through the walls of arteries leading from the heart.
Punnett square	A diagram used to predict the characteristics of offspring produced by two organisms with known combinations of alleles.
pupil	The hole in the front of the eye that light can pass through.
pyramid of biomass	Diagram showing the amount of biomass at different trophic levels of a food chain.
quadrat	A square frame of known area, such as 1 m², that is placed on the ground to get a sample of the organisms living in a small area.

quadriplegia	A condition in which both arms and both legs are paralysed.
qualitative	Qualitative data is in the form of names or descriptions, and not numbers.
quantitative	Quantitative data is data in the form of numbers.
quartiles	When an ordered set of data is divided into four equal groups, those groups are called quartiles.
radioactive	A substance is radioactive if it emits ionising particles or radiation.
radiotherapy	Use of ionising radiation to treat diseases, such as to kill cancer cells.
range	In maths, this is the calculated difference between the highest and lowest values in a set of data, usually ignoring any outliers or anomalous results. In science, we often use the term as a statement of what the highest and lowest values are, without doing the calculation.
rate	How quickly something happens.
ratio	A relationship between two quantities, usually showing the number of times one value is bigger than the other.
receptor cell	A cell that receives a stimulus and converts it into an electrical impulse to be sent to the brain and/or spinal cord.
recessive	Describes an allele that will only affect the phenotype if the other allele is also recessive. It has no effect if the other allele is dominant.
recombinant DNA	DNA made by joining two sections of DNA together.
red blood cell	A biconcave disc containing haemoglobin that gives blood its red colour and carries oxygen around the body to the tissues. Also known as an erythrocyte.
reducing sugars	Small sugar molecules, such as frutose and glucose, that react with Benedict's solution to produce a precipitate.
reflex	A response to a stimulus that does not require processing by the brain. The response is automatic. Also called a reflex action.
reflex arc	A neurone pathway consisting of a sensory neurone passing impulses to a motor neurone, often via a relay neurone, which allows reflexes to occur.
reforestation	Planting new forests where old forests have been cut down.
rejection	When the immune system attacks and kills cells and tissue that come from another person, such as blood (after transfusion) or stem cells.
relay neurone	A short type of neurone found in the spinal cord and brain. Relay neurones link with sensory, motor and other relay neurones.
replicate	When DNA replicates, it makes a copy of itself
resistance	When an organism has resistance to something it is unaffected by it, or not affected very much.
resolution	The smallest change that can be measured by an instrument. For example, in a microscope it is the smallest distance between two points that can be seen as two points and not blurred into one point.
resource	Something that an organism needs to stay alive such as food, water or space.
respiration	A series of reactions occurring in all living cells in which glucose is broken down to release energy.
response	An action that occurs due to a stimulus.
resting metabolic rate	The metabolic rate when the body is at rest.
restriction enzyme	An enzyme that cuts DNA molecules into pieces
retina	The part at the back of the eye that changes energy transferred by light into nerve impulses. Contains rods and cones.
ribosome	A sub-cellular structure that attaches to mRNA. It allows tRNA molecules to match up with the mRNA codons and also joins the amino acids together.
RNA	Abbreviation for ribonucleic acid. The molecule is made of phosphate groups and ribose sugars linked together with one of four bases.
RNA polymerase	An enzyme that creates mRNA from DNA.
rod (cell)	A cell in the retina that detects low levels of light. It cannot detect different colours.
root hair cell	A cell found on the surface of plant roots that has a large surface area to absorb water and dissolved mineral salts quickly from the soil.
rooting powder	A powder containing plant hormones into which cuttings are dipped to speed up the growth of new roots.
sample	A small portion of an area or population.
Sankey diagram	A diagram showing energy transfers, where the width of each arrow is proportional to the amount of energy it represents.
scale bar	A line drawn on a magnified image that shows a certain distance at that magnification.
scientific paper	An article written by scientists and published in a science magazine called a journal. It is like an investigation report, but usually shows the results and conclusions drawn from many experiments.
screening	Tests on samples of body fluids to check if people have a certain condition, e.g. an STI.
secondary consumers	Carnivores, that eat primary consumers (herbivores).
secondary infection	An infection due to the immune system being weakened previously by a different pathogen.
secondary response	The way in which the immune system responds on the second occasion that a particular pathogen enters the body.
selective breeding	When humans choose an organism that has a certain characteristic and breed more of these organisms, making that chosen characteristic more and more obvious.

selective reabsorption	Taking back particular (useful) substances from a nephron, such as glucose and some mineral ions.
selective weedkiller	A substance that kills a certain type of plant only, leaving others unaffected.
semi-permeable	Another term for partially permeable.
semi-quantitative	Qualitative data that gives an idea of order e.g. small, bigger, biggest.
sense organ	An organ that contains receptor cells
sensory neurone	A neurone that carries impulses from receptor cells towards the central nervous system.
septum	The dividing wall between the left and right atria and ventricles in the heart.
sewage	Human waste collected for treatment.
sex chromosome	A chromosome that determines the sex of an organism.
sex hormone	Any hormone that affects reproduction.
sex-linked genetic disorder	A disorder caused by genes that is inherited differently in males and females, such as red–green colour blindness, which is more common in men than in women.
sex organ	An organ in the reproductive system.
sexual intercourse	Sexual activity between two people, especially when the penis is placed in the vagina.
sexual reproduction	Reproduction that needs a male and female parent.
sexually transmitted infection (STI)	A communicable disease that can be passed from an infected person to an uninfected person during sexual activity.
shivering	Rapid contraction and relaxation of muscles that causes the body to warm up.
short-sightedness	An eye condition in which distant objects appear blurred.
side effect	Unintended (often unpleasant or harmful) effect of a medicine
sieve tubes (and cells)	Tubes formed of plant phloem sieve cells (so called because the cells have holes in their ends). The tubes carry sugars and other compounds around the plant.
solute	Describes a substance that dissolves in a liquid to make a solution
solvent	Describes the liquid in which a substance dissolves to make a solution
specialised cell	A cell that is adapted for a certain specific function (job).
species	A group of organisms that can reproduce with each other to produce offspring that will also be able to reproduce.
specific	A particular requirement e.g. an enzyme is specific and only reacts with one particular substrate.
sperm cell	The male gamete (sex cell).
spinal cord	The large bundle of nerves leading from the brain and down the back.
spindle fibre	A filament formed in a cell during mitosis, which helps to separate chromosomes.
spongy cells	Irregularly shaped cells in a plant leaf that form air spaces between them.
stain	A dye used to colour parts of a cell to make them easier to see.
standard form	A very large or very small number written as a number between 1 and 10 multiplied by a power of 10. Example: $A \times 10^n$ where A is between 1 and 10 and n is the power of 10.
stem cell	An unspecialised cell that continues to divide by mitosis to produce more stem cells and other cells that differentiate into specialised cells.
stent	A small mesh tube used to widen narrowed blood vessels and allow blood to flow more easily.
sterilise	To kill all microorganisms on or in something.
sticky end	A short section of single-stranded DNA at the end of a piece of DNA that has been cut by a restriction enzyme.
stimulus	A change in a factor (inside or outside the body) that is detected by receptors. Plural is stimuli. Examples include light and sound.
stoma	A tiny pore in a leaf which, when open, allows gases to diffuse into and out of the leaf. Plural is stomata.
storage organ	A plant organ used to store energy-rich substances such as starch. For example, a potato.
strain	Bacteria of a species that are slightly different from other strains of the species.
stroke	Death of brain cells caused by a lack of blood, due to a blockage in a blood vessel in the brain.
stroke volume	The volume of blood the heart can pump out with each beat.
substrate	A substance that is changed during a reaction.
sucrose	The type of sugar found in the phloem of plants, and used as table sugar.
surface area to volume ratio (SA : V)	The total amount of surface area of an object divided by its volume.
sustainability	Ability to continue something, such as food production, at the same level without negative effects now or in the future.
symptom	Something that is suffered when an organism is ill (e.g. pain) or is a sign of illness (e.g. a high temperature).
synapse	The point at which two neurones meet. There is a tiny gap between neurones at a synapse.
synthesis	To build a large molecule from smaller subunits.
target organ	An organ on which a hormone has an effect.
telophase	The stage of mitosis in which the chromosomes arrive at opposite ends of the cell and the nucleus membrane reforms.

template strand	The strand of a DNA molecule that RNA polymerase uses to make mRNA.
tendon	A cord of tissue that connects a muscle to a bone.
testis	An organ in the male reproductive system that produces sperm cells and the hormone testosterone. Plural is testes.
thermoregulation	The control of body temperature, especially in core parts of the body (e.g. heart, liver and brain).
thymine	One of the four bases found in DNA. Often written as T and pairs up with adenine.
thyroid gland	A gland that releases the hormone thyroxine into the blood. Can be referred to as 'the thyroid'.
thyroxine	A hormone released by the thyroid gland that affects metabolic rate by changing how certain cells work, e.g. causes heart cells to contract more strongly.
tissue culture	Growing tiny pieces of tissue, or cells, in the lab.
transcription	The process by which the genetic code in one strand of DNA molecules is used to make mRNA.
transfer RNA (tRNA)	A molecule of RNA that carries an amino acid.
translation	The process by which the genetic code in a molecule of mRNA is used to make a polypeptide.
translocation	The transport of sugars (mainly sucrose) and other compounds in the phloem tissue of a plant.
transpiration	The flow of water into a root, up the stem and out of the leaves.
trophic level	Feeding level in a food chain, such as producer, primary consumer, secondary consumer.
tropism	A response to a stimulus in which an organism grows towards or away from the stimulus. A positive tropism is a growth towards a stimulus and a negative tropism is a growth away from the stimulus.
tuberculosis (TB)	A communicable bacterial disease that infects the lungs.
tumour	A lump formed of cancer cells
type 1 diabetes	A type of diabetes in which the pancreas does not produce insulin.
type 2 diabetes	A type of diabetes in which cells do not respond to insulin, or too little insulin is produced.
ulcer	A sore area in the stomach lining
uracil (U)	A base found in RNA but not in DNA.
urea	A waste product produced in the liver from excess amino acids.
urinary system	Organ system that removes excess substances and waste products from the body in urine.
urine	A fluid produced by the kidneys, containing urea and other waste or excess substances dissolved in water.
vaccine	A substance containing dead or weakened pathogens (or parts of them), introduced into the body to make a person immune to that pathogen.
vacuole	The membrane-bound space in the cytoplasm of cells. Plant cells have a large permanent vacuole, which stores water and nutrients, and helps to support the plant by keeping the cells rigid.
vasoconstriction	Narrowing of blood vessels, which reduces blood flow.
vasodilation	Widening of blood vessels, which increases blood flow.
valve	Flaps of tissue in the heart and veins that close to make sure blood only flows in one direction.
variation	Differences in the characteristics of organisms.
variety	Group of plants of the same species that have characteristics that make them different to other members of the species.
vector (disease)	Something that transfers things from one place to another. For example, an organism that carries a pathogen from one infected person to another, such as the mosquito that carries the malaria protist.
vein	A blood vessel that transports blood towards the heart.
vena cava	A major vein leading to the heart.
ventricle	A lower chamber in the heart that pumps blood out into arteries.
vertebrate	Animal with bones, such as a human.
virus	A particle that can infect cells and cause the cells to make copies of the virus.
waist : hip ratio	A measure of the amount of the fat in the body, calculated by dividing the waist circumference by the hip circumference.
water cycle	A sequence of processes by which water moves through the abiotic and biotic parts of an ecosystem.
weeds	Plants that cause problems, such as by competing with crop plants for light, water and nutrients.
white blood cell	A type of blood cell that forms part of the body's defence system against disease.
wilt	Drooping of parts of a plant caused by a lack of water.
xylem vessel (and cells)	A long, thick-walled tube found in plants, formed from many dead xylem cells. The vessels carry water and dissolved mineral salts through the plant.
yield	The amount of useful product that you can get from something.
zygote	A fertilised egg cell.

Index